PLASMA DISCHARGE in LIQUID

Water Treatment and Applications

PLASMA DISCHARGE in LIQUID

Water Treatment and Applications

Yong Yang
Young I. Cho
Alexander Fridman

CRC Press
Taylor & Francis Group
Boca Raton   London   New York

CRC Press is an imprint of the
Taylor & Francis Group, an **informa** business

CRC Press
Taylor & Francis Group
6000 Broken Sound Parkway NW, Suite 300
Boca Raton, FL 33487-2742

First issued in paperback 2017

Version Date: 2011922

ISBN 13: 978-1-4398-6623-8 (hbk)
ISBN 13: 978-1-138-07495-8 (pbk)

Library of Congress Cataloging-in-Publication Data

Yang, Yong.
   Plasma discharge in liquid : water treatment and applications / Yong Yang and Young L. Cho.
       p. cm.
   Includes bibliographical references and index.
   ISBN 978-1-4398-6623-8
   1. Water--Purification. 2. Ionization of gases. 3. Plasma dynamics. I. Cho, Young L. II. Title.

TD437.Y36 2012
628.1'62--dc23   2011032561

Visit the Taylor & Francis Web site at
http://www.taylorandfrancis.com

and the CRC Press Web site at
http://www.crcpress.com

Contents

Preface

Plasma plays an important role in a wide variety of industrial applications, including material processing, semiconductor manufacturing, light sources, propulsion, and many more. As a branch of plasma physics, plasmas in liquids were historically studied by the electrical engineering community for pulsed power applications and high-voltage insulation. Liquids, particularly water, usually have higher dielectric constants and higher dielectric strengths than gas phases.

Interest has increased recently in plasma discharges in liquids because of the potential applications for various biological, environmental, and medical technologies. For example, electric breakdown is developed as a nonchemical method for biofouling removal and contaminant abatement in water, with a potential for extension into a wide range of other water treatment applications. Plasma methods that effectively combine the contributions of ultraviolet (UV) radiation, active chemicals, and high electric fields can be considered as an alternative to conventional water treatment methods. However, knowledge of the electric breakdown of liquids has not kept pace with this increasing interest, mostly due to the complexity of phenomena related to the plasma breakdown process.

The motivation of this book was to provide engineers and scientists with a fundamental understanding of the physical and chemical phenomena associated with plasma discharges in liquids, particularly in water. This text has been organized into two parts. Part 1 addresses the basic physics of electric breakdown in liquids. Chapter 2 examines the generation of plasma in liquids, while Chapter 3 provides an introduction to the elementary processes of the plasma initiation mechanism based on both electronic and bubble theories. Part 2 addresses various applications of underwater plasma discharges in the water treatment industry, specifically examining plasma-assisted volatile organic compound decontamination and remediation of contaminated water (Chapter 4), microorganism sterilization and other biological applications (Chapter 5), and cooling water treatment (Chapter 6). We have drawn on extensive work in recent publications on the aforementioned subjects, and they believe that this book will serve as a valued reference for those who are interested in these topics.

We acknowledge the research support provided by the U.S. Department of Energy, National Energy Technology Laboratory. In addition, we are grateful to colleagues and friends from A. J. Drexel Plasma Institute, especially Dr. Gary Friedman, Dr. Greg Fridman, Dr. Alex Rabinovich, Dr. Alex Gutsol, and Dr. Andrey Starikovskiy, for their stimulating discussions on the topic of plasma and immeasurable assistance in the preparation and development of the book.

About the Authors

Dr. Young I. Cho has been a professor at Drexel University in Philadelphia since 1985. Prior to joining Drexel University, he spent four years at NASA's Jet Propulsion Laboratory, California Institute of Technology, as a member of the technical staff. His research interest includes fouling prevention in heat exchangers, physical water treatment using electromagnetic fields, hemorheology, and energy. Currently, he is developing methods of applying low-temperature plasma technology to prevent mineral and biofouling problems in cooling water. He has authored or coauthored approximately 250 papers in the area of heat transfer, fluid mechanics, rheology, and energy. He was an editor for *Handbook of Heat Transfer* (McGraw Hill, 3rd ed.) and *Advances in Heat Transfer* (Academic Press). He was the recipient of the 1992 Lindback Award for excellence in teaching at Drexel University. In 1993, Dr. Cho was the chairman of the Advanced Fluid Committee under the International Energy Agency. In 1995, he was the recipient of the Research Professor of the Year at Drexel University. He received his PhD from the University of Illinois, Chicago, in 1980.

Dr. Alexander Fridman is Nyheim Chair Professor at Drexel University, Philadelphia, and director of the A. J. Drexel Plasma Institute. He develops novel plasma approaches to material treatment, fuel conversion, hydrogen production, aerospace engineering, biology, and environmental control. Recently, significant efforts of Dr. Fridman and his group have been directed to development of plasma medicine, which is a revolutionary breakthrough area of research focused on direct plasma interaction with living tissues and direct plasma application for wound treatment, skin sterilization, blood coagulation, and treatment of different diseases, not previously effectively treated.

Dr. Fridman worked and taught as a professor and researcher in different national laboratories and universities in the United States, France, and Russia. He has had 7 books and more than 550 scientific papers published and has chaired several international plasma conferences. Dr. Fridman has received numerous awards, including International Plasma Medicine Award, Stanley Kaplan Distinguished Professorship in Chemical Kinetics and Energy Systems, George Soros Distinguished Professorship in Physics, the DuPont research award, Chernobyl award, University of Illinois and Drexel Research awards, and Kurchatov Medal for Scientific Achievements. Dr. Fridman, together with the Nobel Prize laureate N. G. Basov, received the State Prize of the Soviet Union for discovery of selective stimulation of chemical processes in nonthermal plasma.

Dr. Yong Yang has been an associate professor at the College of Electrical and Electronic Engineering, Huazhong University of Science and Technology (HUST) in Wuhan, China, since 2011. Prior to joining HUST, he spent five years at Drexel Plasma Institute, Drexel University, pursuing his PhD degree. His research interests include low-temperature plasma discharges in liquid and atmospheric gas and their applications in environmental, medical, and energy-related fields. He has authored or coauthored over 20 scientific papers in the area of heat transfer and low-temperature plasma. He was the recipient of the Provost Fellowship, George Hill Fellowship, and 2011 Research Excellence Award at Drexel University. He received his BS and MS from Tsinghua University in Beijing, China, in 2003 and 2006, respectively, and his PhD from Drexel University in Philadelphia in 2011.

1

Introduction

1.1 Background

Plasma is often referred to as the fourth state of matter in which a certain portion of particles in gas or liquid is ionized. The term *plasma* was first introduced by Irving Langmuir; the way an electrified fluid carried electrons and ions moving at high velocity reminded him of the way blood plasma carried red and white corpuscles. In his article published in the *Proceedings of the National Academy of Sciences* in 1928, he wrote: "Except near the electrodes, where there are sheaths containing very few electrons, the ionized gas contains ions and electrons in about equal numbers so that the resultant space charge is very small. We shall use the name plasma to describe this region containing balanced charges of ions and electrons."

The ionization of the neutral particles is usually achieved through heating. As temperature rises, molecules become more energetic and transform in sequence from solid to liquid, gas, and a plasma state. In the plasma state, freely moving particles, including electrons and positively or negatively charged ions, make them electrically conductive and can attain electrical conductivities sometimes larger than metals such as gold and copper.

1.2 Plasma Generation in Nature and in the Laboratory

Plasmas comprise the majority of matter in the universe. Most of the stars are made of plasma. The space between the stars is filled with plasma, although at a much lower density than that inside the stars. On Earth, however, naturally occurring plasma is somewhat rare. In Earth's atmosphere, the best-known plasma phenomenon is lightning. An average lightning bolt carries an electric current of about 100 kA and an approximate power output of 1 MW per meter, which rapidly heats the air in its immediate vicinity to a temperature of over 10,000°C. The sudden heating effect and the expansion

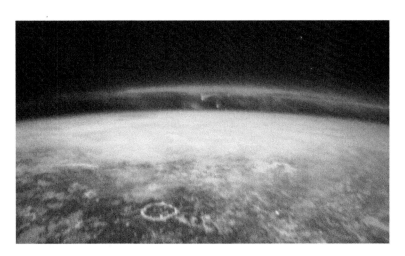

FIGURE 1.1
(See color insert.) Aurora borealis as seen from International Space Station. (Courtesy of NASA.)

of heated air give rise to a supersonic shock wave in the surrounding clear air. Once this shock wave decays to an acoustic wave, it is heard as thunder.

At an altitude of approximately 100 km, the atmosphere is conductive due to the ionization of neutral molecules by solar radiation, making this region of the atmosphere in a plasma state called the *ionosphere*. Long-distance communication is largely made possible by the presence of the ionosphere through the reflection of radio waves by the ionized layer. Aurora is another example of natural plasma on Earth (see Figure 1.1). At near-space altitudes, Earth's magnetic field interacts with charged particles from the Sun. These particles are diverted and often trapped by the magnetic field. These trapped particles are most dense near the poles, causing ionizations of neutral particles in the atmosphere and thus accounting for the light emission of the aurora.

Although the presence of natural plasma on Earth is relatively scarce (Figure 1.2), the number of industrial applications of plasma technologies is extensive. Historically, the study of vacuum tubes and so-called cathode rays laid the initial foundation of much of our understanding of plasma, which led to the development of plasma lighting technologies since the 19th century. More energy-efficient fluorescent lamps have been available on the market for the past few decades. In recent years, high-output radio-frequency (RF)-powered lamps have been developed as a viable alternative to LED (light-emitting diode) lamps, whose manufacturing process also heavily relies on plasma technologies.

Another important application of plasma resides in the semiconductor manufacturing industry. The microelectronics industry would virtually be impossible without plasma since most processes in semiconductor device fabrication, including dry etching, deposition, and implantation, cannot be achieved by any other commercial method but plasma.

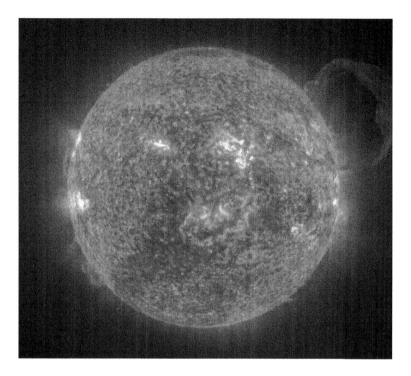

FIGURE 1.2
(See color insert.) Solar plasma. Emission in spectral lines shows the upper chromosphere at a temperature of about 60,000 K. (Courtesy of NASA.)

Plasma is widely employed in the coating industry, in which its large enthalpy content, high temperature, and high deposition rates are advantageous for increased throughputs. Various materials, including plastics, complex alloys, composites, and ceramics, can be deposited over a large area in different shapes. In the plasma-spraying process (see Figure 1.3), the material to be deposited—typically in a powder form—is introduced into a plasma jet with a temperature on the order of 10,000 K. The material is melted and accelerated toward the surface of the substrate, where the molten droplets rapidly solidify and form the deposition layer.

Surface property modifications for different polymer materials are usually performed using plasma. Many common polymer surfaces are chemically inert and therefore pose challenges for use as substrates for applied layers. The modification of polymer surfaces by plasma treatment can improve surface characteristics such as adhesion promotion, enhancement of wettability and spreading, improved biocompatibility, functionalized surface, reduced surface friction, and tackiness. These unique surface modifications that can be achieved using the plasma process result from the effects of the photons and active species in the plasma to react with surfaces in depths from several hundred angstroms to microns without influencing the bulk properties of the polymer base material.

FIGURE 1.3
Plasma thermal spray coating. (Courtesy of NASA.)

Low-temperature, nonequilibrium plasmas are an emerging technology for abating volatile organic compound (VOC) emissions and other industrial exhausts, which have become an important environmental concern as most of them are carcinogens and harmful to living organisms. Abatement of these polluting substances is conventionally handled by water scrubbers or adsorbent filters to convert them to harmless products. However, for the abatement of diluted VOCs with low concentrations (<100 ppm), these conventional techniques are not suitable, mainly due to high-energy consumption. Among the alternatives, nonequilibrium atmospheric pressure plasma processes have been shown to be effective in treating a wide range of emissions, including aliphatic hydrocarbons, chlorofluorocarbons, methyl cyanide, phosgene, as well as sulfur and nitrogen oxides. The reduction of the power consumption relies on the selective production of reactive species like ions, radicals, and activated molecules by the plasma process without heating of the bulk volume.

1.3 Needs for Plasma Water Treatment

The availability of clean water is an issue that has paralleled the continual increase in water consumption due to both global population growth and

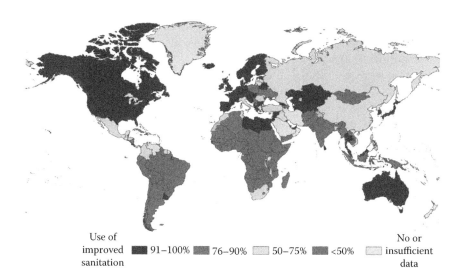

FIGURE 1.4
(See color insert.) Use of sanitary water in rural areas, 2008. (From World Health Organization. (2010) Progress on sanitation and drinking-water.)

the economic development in a number of developing countries. From a global perspective, an estimated 2.6 billion people are unable to acquire clean, safe drinking water (World Water Assessment Programme, 2009). The global picture shows great disparities between regions (Figure 1.4). Virtually the entire population of the developed regions uses improved facilities for water supply and discharge, but in developing regions only around half the population use improved sanitation facilities.

Contaminated water can be attributed to a number of factors, including chemical fouling, inadequate treatment, and a deficient or failing water treatment and distribution system. An additional important cause of the contamination is the presence of untreated bacteria and viruses within the water. The United Nations World Health Organization (WHO; 2010) estimated that nearly 35% of all deaths in developing countries were related directly to contaminated water. In the United States, the increased presence of *Escherichia coli* (*E. coli*) along with various other bacteria within some areas has also become a cause for national concern. In 2006, there was an outbreak of disease caused by *E. coli* found in spinach in 25 U.S. states, which caused thousands of illnesses and three deaths (U.S. Environmental Protection Agency Office of Water, 2009). In 2010, more than 500 million eggs were recalled after dangerous levels of *Salmonella* were detected. Salmonella may be caused by groundwater that has been contaminated by animal feces.

In an effort to inactivate these bacteria, traditional chemical treatments, ultraviolet (UV) radiation, and ozone injection units have been implemented for potable water delivery systems. The experimental success and commercialization of these water treatment methods are not, however, without deficiencies. With regard to human consumption of water, chemical treatments such as chlorination can render potable water toxic. Both UV radiation and ozone injection have been proven to be practical methods of bacterial inactivation in water, but the effectiveness of such methods largely depends on adherence to regimented maintenance schedules.

Plasma methods that effectively combine the contribution of UV radiation, active chemicals, and high electric fields have been considered as an alternative to these conventional water treatment methods (Locke, Sato, et al., 2006; Fridman, Gutsol, and Cho, 2007; Muhammad, 2010). Before considering direct application of plasma to water treatment (which is a major goal of this book), we discuss briefly the independent application of UV radiation, active chemicals, and high electric fields for the deactivation of microorganisms in water.

1.4 Conventional Water Treatment Technologies

Currently, there are many available methods of water treatment and decontamination, including chlorination, ozonation, UV radiation, in-line filters, and pulsed electric fields. Many of these systems are utilized in large industrial applications. However, methods such as chlorination, in-line filtering, and UV radiation are also applied in point-of-use applications, including treatment of swimming pool and well water. These methods have distinct advantages and disadvantages and are carefully analyzed next.

1.4.1 Chlorination

The technique of purification of water using chlorine was first proposed in the early 1800s. For the past 200 years, chlorine has remained both an acceptable and a widely employed method of treatment with regard to water disinfection due to its ease of use and associated efficiency for the inactivation of microorganisms. Regardless of the system size, it is one of the least-expensive disinfection methods. However, the chlorination of public drinking water supplies is meeting with strong resistance as people are more concerned about the health effects of the process as the toxicity of chlorine requires strict adherence to accepted concentration levels. An excess of chlorine in a drinking water supply could render the water toxic with regard to human ingestion. Unwanted disinfection by-products (DBPs) resulting from the interaction of chlorine with other chemicals present in water can prove corrosive and deteriorative to the system. Under some circumstances, chlorine can react with organic compounds found in the water supply to produce trihalomethanes (THMs) and haloacetic acids (HAAs) (Adams et al., 2005), both of which are highly carcinogenic. In addition, because a chlorination-based system must be continually replenished, the storage and transportation of this chemical becomes a significant hazard.

1.4.2 In-Line Filters

In-line filters are commonly used to remove undesirable substances from water. Many different types are commercially available, including activated filters, microfilters, and reverse osmosis filters. The key advantage to these filters is that they do not require power to operate, but there are two significant drawbacks to this method. Although these filters are capable of preventing microorganisms from passing through the system, they are incapable of inactivating them, resulting in bacterial growth in the filters. The small pores needed to trap microorganisms also inhibit the flow, resulting in pressure loss across the filter. Significant pressure losses in the system require a larger-size pump.

1.4.3 Pulsed Electric Field

The next method considered for inactivating microorganisms is pulsed electric field technology. Since the electric field associated with this method is not strong enough (membrane potential of more than 1 V can kill a bacterium) to initiate electrical breakdown in water, there is no resulting electric discharge. The deactivation of microorganisms is believed to be due to electroporation, a process that is the creation of holes in cell membranes, indicating that plasma-originated electric fields (for example, those in DBD streamers) might be sufficient. At nominal conditions, the energy expense for a two-log reduction is approximately 30,000 J/L (Katsuki et al., 2002). Researchers at the Technical University of Hamburg, Germany, reported pulsed electric field effects on suspensions of bacteria in water (Grahl and Markl, 1996). They reported that the external electric pulse produced a membrane potential of more than 1 V for the effective killing of bacteria.

1.4.4 Ultraviolet Radiation

Ultraviolet radiation generated by plasma has proven effective in decontamination processes and is gaining popularity, particularly in Europe, because it does not leave undesirable by-products in water. Measurement of this radiation is considered in terms of dosage, which is given as the product of intensity (watts per square centimeter, W/cm^2) and contact time (seconds, s). Most bacteria and viruses require relatively low UV dosages for inactivation, which is usually in a range of 2,000–6,000 mW·s/cm^2 for a 90% kill rate. For example, *E. coli* requires a dosage of 3,000 mW·s/cm^2 for a 90% reduction (Wolfe, 1990). Cryptosporidium, which shows an extreme resistance to chlorine, requires a UV dosage greater than 82,000 mW·s/cm^2. The criteria for the acceptability of UV disinfecting units include a minimum dosage of 16,000 mW·s/cm^2 and a maximum water penetration depth of approximately 7.5 cm (Wolfe, 1990).

UV radiation in the wavelength range from 240 to 280 nm causes irreparable damage to the nucleic acid of microorganisms. The most potent wavelength of UV radiation for DNA damage is approximately 260 nm. Currently, there are two types of commercial UV lamps: low-pressure and medium-pressure mercury lamps. It is worth noting that the UV radiation from both types is generated in plasma. The former possesses a relatively low temperature and produces a narrow band of UV light with a peak near the 254-nm wavelength, whereas the latter produces a higher temperature and a broader band of UV and has a much greater treatment capacity, approximately 25 times higher than the former (Wolfe, 1990). The life of a UV lamp is relatively short, approximately 8,000–10,000 h, compromised by several additional factors, including biological shielding and chemical or biological film buildup on the surface of the lamp. An advantage of this system is that both the temperature and the pH of the treated water are not significantly affected, and no undesirable by-products are created (Wolfe, 1990). However, the total energy cost of the UV water treatment is high, similar to that for pulsed electric fields.

The UV photons can have two possible effects on a microorganism. One effect is through direct collisions with contaminants, causing the mutation of bacterial DNA. This prevents proper cellular reproduction and thus effectively inactivates the microorganism. Alternatively, the photons can provide the necessary energy to ionize or dissociate water molecules, thus generating active chemical species. Both mechanisms increase the deactivation of viable microorganisms (Sun et al., 2006). It has been suggested that the UV system produces charged particles in water such that charge accumulation occurs on the outer surface of the bacterial cell membrane. Subsequently, the electrostatic force overcomes the tensile strength of the cell membrane, causing its rupture at a point of small local curvature as the electrostatic force is inversely proportional to the local radius squared. Note that since the membrane of gram-negative bacteria such as *E. coli* often possesses irregular surfaces, UV disinfection becomes more effective for gram-negative bacteria than gram-positive ones (Laroussi, 2005; Laroussi et al., 2002; Hurst, 2005).

Researchers at Macquarie University, Australia, studied new UV light sources for the disinfection of both drinking water and recycled wastewater (Carman et al., 2003). They reported that UV lamps were much more effective than chlorine in dealing with the hundreds of potentially dangerous types of microbes in water, including the well-known giardia and cryptosporidium. The UV radiation did not blow the microbe apart as such. Instead, it entered through the outer membrane of the bacteria into the nucleus and actually cut the bonds of the DNA so that the bacteria could not repair themselves and could not reproduce.

1.4.5 Ozonation

Ozonation is a growing method of water treatment; compressed ozone gas is bubbled into a contaminated solution and dissolves in it. The two most common methods of generating ozone utilize plasma: corona discharge and dielectric barrier discharge. The ozone (O_3) is one of the most well-known active chemical species and is capable of efficiently inactivating microorganisms at a level comparable to chlorine. The residence time of the ozone molecules in the solution depends on temperature. At high temperatures, ozone decomposition to molecular oxygen takes place quickly. Solutions maintained at low temperatures tend to have faster deactivation times when compared to solutions maintained at high temperatures. Achieving a four-log reduction at 20°C with an ozone concentration of 0.16 mg/L requires an exposure time of 0.1 min (Anpilov et al., 2001). At higher temperatures and pH levels, ozone tends to rapidly decay and requires more exposure time. Due to the corrosive and toxic nature of ozone, ozonation systems require a high level of maintenance.

Plasma discharge, especially dielectric barrier discharges (DBD), has been used for the production of ozone for the past several decades for water treatment purposes. Ozone has a lifetime of approximately 10–60 min, which varies depending on the pressure, temperature, and humidity of surrounding conditions. Because of the relatively long lifetime of ozone, ozone gas is remotely produced in air or oxygen, stored in a tank, and injected into water using a compressor. Of note is that hydrogen peroxide is also produced when ozone is produced in a plasma discharge in humid air. However, the half-life of the hydrogen peroxide is much shorter, so it could not be directly used for conventional water treatment systems.

The feasibility of using ozonation also was tested for the ballast water treatment for large ships. Dragsund, Andersen, and Johannessen (2001) reported *Ct* values for various organisms. Note that the *Ct* value is defined as the product of ozone concentration *C* (mg/L) and the required contact time *t* (min) to disinfect a microorganism in water. For example, for *Ditylum brightwelli* (important ballast water species), the *Ct* value was 50 mg·min/L. In other words, if the ozone concentration is 2 mg/L, it takes 25 min of contact time to disinfect this organism in ballast water. They reported that ozone reacted with seawater and produced a number of corrosive compounds (mostly compounds of chlorine). The long contact time between ozone and organisms is beneficial for the disinfection of organisms but harmful in producing corrosion of the ballast tank. However, the half-life of ozone is relatively short compared to the time required for corrosion, such that the corrosion threat may not last long. One of the reasons why ozone has not been used widely for water treatment in the United States is the relatively high cost of producing ozone, a process that requires dry air or a concentrated oxygen supply, compressor, ozone gas injection system, and electricity. Furthermore, if ozone gas is accumulated in a closed space by accident, it can be highly toxic to humans. In addition, the energy efficiency of ozonation is limited by O_3 losses during storage and transportation.

In summary, ozone and UV radiation generated in remote plasma sources are effective means of water cleaning and sterilization. If plasma is organized not remotely but directly in water, the effectiveness of the treatment due to plasma-generated UV radiation and active chemical species can be much higher. The organization of plasma inside water also leads to an additional significant contribution of short-living active species (electronically excited molecules, active radicals like OH, O, etc.), charged particles, and plasma-related strong electric fields to cleaning and sterilization (Sun, Sato, & Clements, 1997; Locke, Sato, et al., 2006; Fridman, Gutsol, & Cho, 2007; Sun, Kunitomo & Igarashi, 2006). While direct water treatment by plasma generated in water can be effective, both the initiation and sustaining of plasma in water (where the mean free path of electrons is very short) are more complicated than in the gas phase, a subject discussed in the following sections.

1.5 Plasma in Liquids

Historically, plasmas in liquids were studied by the electrical engineering community for pulsed power applications and high-voltage insulation (Figure 1.5). Liquids, particularly water, usually have a higher dielectric constant and dielectric strength than gas phases and have been widely used as an insulating media for high-voltage pulse lines in pulsed power systems. For

FIGURE 1.5
(See color insert.) Sandia National Lab's Z machine, bathed in transformer oil and deionized water for greater electric insulation. (Courtesy of Sandia National Lab, 2004.)

example, high molecular weight hydrocarbons are frequently used in liquid-filled transformers for both insulation and cooling purposes. The conduction or insulation behavior of liquid is determined by the Maxwellian relaxation time, which is the ratio of dielectric permittivity and electric conductivity. Pure water has a relative dielectric constant e_r of 80 up into the gigahertz range, and its electrical conductivity s is usually a few microsiemens per centimeter, resulting in its Maxwellian relaxation time on the order of a few microseconds. Given that a specific water is exposed to an electric pulse with a long duration time of Dt, that is, when $Dt \gg e_r e_0/s$, where e_0 is vacuum permittivity, the aqueous solution behaves as a resistive medium. One of the major results of such a long electric pulse is the electrolysis of water with the production of hydrogen and oxygen. For much shorter times, that is, when $Dt \ll e_r e_0/s$, water behaves as a dielectric medium and can sustain a high electric stress until a breakdown threshold is reached. Critical electric breakdown fields of megavolts per centimeter have been reported (Locke, Sato, et al., 2006). These numbers indicate that high energy densities could be achieved in water. The high dielectric strength of liquids allows considerably higher currents, thus reducing the size of high-power switches compared to gaseous switches. Besides the advantage of a lower switch inductance, the higher recombination and diffusion rate due to the higher density of liquids will quickly restore the switching medium. Small switching volumes also allow liquid removal quickly after each shot and operation at high repetition rates. However, this type of switching is not without drawbacks, one of which is shot-to-shot timing instability (jitter) originating from the stochastic nature of the breakdown processes. This jitter can be significant, making liquid switches not suitable for accurate timing and synchronization purposes.

There is increasing interest in plasma discharge in liquid, mostly because of its potential applications for various biological, environmental, and medical technologies. For example, electric breakdown has been developed as a nonchemical method for biofouling removal and contaminant abatement in water, with a potential for extension into a wide range of other water treatment applications. The simultaneous production of intense UV radiation, shock waves, and various chemical products, including OH, O, HO_2, and H_2O_2 from the electric breakdown in water, is utilized (Chu et al., 2006). The synergetic effects of these products are believed to produce higher efficiencies than traditional water treatment technologies utilizing each product separately. Another application is to use plasma as an etching tool for the removal of biological tissues. A plasma scalpel in saline solution is able to etch flesh and clean wounds during surgery and is close to widespread practical use (Stalder, McMillen, and Woloszko, 2005). Shock waves produced by high-energy plasma discharges inside liquids are used for various applications, including underwater explosions (Akiyama, Sakugawa, T. and Namihira, 2007), rock fragmentation (Bluhm, Frey, and Giese, 2000), and lithotripsy (Sunka, 2001). In these applications, it is important to understand the mechanism and dynamics of the electric breakdown process in liquids, a subject of the present book that has been under investigation for more than 100 years, with a number of experimental results reported.

1.5.1 Mechanisms of Plasma Discharges in Liquids

Mechanisms of plasma discharges and breakdowns in liquids (specifically in water) can be classified into two groups: The first group considers the breakdown in water as a sequence of a bubble process and an electronic process within the bubbles, while the second group divides the process into a partial discharge and a fully developed discharge, such as arc or spark (Akiyama, 2000). According to the approach from the first group, the bubble process starts from a microbubble formed by the vaporization of liquid from local heating in the strong electric field region at the tip of an electrode. As the bubble grows, an electrical breakdown subsequently takes place within the bubble. In this case, a cavitation mechanism was suggested to explain the slow bushlike streamers (Beroual, 1993; Beroual, Zahn, and Badent, 1998). The appearance of bright spots is delayed from the onset of the application of high voltage, and the delay time tends to be greater for low applied voltages. The time lag to water breakdown increases with increasing pressure, supporting the bubble mechanism in a submicrosecond discharge formation in water (Jones & Kunhardt, 1994, 1995). The time to form the bubbles was about 3–15 ns, depending on the electric field and pressure (Akiyama, 2000). The influence of the water's electrical conductivity on this regime of the discharges was small (Akiyama, 2000).

Bulk heating via ionic current does not contribute to the initiation of the breakdown. The power necessary to evaporate water during streamer propagation can be estimated using the streamer velocity, the size of the streamer, and the heat of vaporization (Lisitsyn et al., 1999a). Using a streamer radius of 31.6 mm, a power of 2.17 kW was estimated to be released into a single streamer to ensure its propagation in the form of vapor channels. In the case of multiple streamers, the required power can be estimated by multiplying the number of visible streamers to the power calculated for a single streamer. In the electronic process, both electron injection and drift in liquid take place at the cathode, while hole injection through a resonance tunneling mechanism occurs at the anode (Katsuki et al., 2002). In the electronic process, breakdown occurs when an electron makes a suitable number of ionizing collisions in its transit across the breakdown gap.

According to the approach in the second group on the mechanisms of electrical discharges in water, the discharge process is divided into partial electrical discharges and arc or spark discharge (Locke, Sato, et al., 2006; Sato, Ohgiyama, & Clements, 1996; Sugiarto, Ohshima, & Sato, 2002; Sun, Sato, & Clements, 1999; Sugiarto et al., 2003; Manolache, Shamamian, & Denes, 2004; Ching et al., 2001; Ching, Colussi, & Hoffmann, 2003; Robinson, Ham, & Balaster, 1973; Robinson, Ham, and Balaster, 1973). In the partial discharges, the current is mostly transferred by ions. For the case of water with a high electric conductivity, a large discharge current flows, resulting in a shortening of the streamer length due to the faster compensation of the space charge electric fields on the head of the streamer. Subsequently, a higher power density in the channel is obtained, resulting in a higher plasma temperature, higher UV radiation, and the generation of acoustic waves. In the arc or spark discharge, the current is transferred by electrons. The high current heats a small volume of plasma in the gap between the two electrodes, generating a quasithermal plasma. When a high-voltage, high-current discharge takes place between two submerged electrodes, a large part of the energy is consumed in the formation of a thermal plasma channel. This channel emits UV radiation, and its expansion against the surrounding water generates intense shock waves (Sunka et al., 1999; Lee et al., 2003).

In 2006, Locke, Sato, et al. published a comprehensive review of the application of strong electric fields in water and organic liquids; they included 410 references. They explained in detail the types of discharges used for water treatment, physics of the discharge, and chemical reactions involved in the discharge in water. Bruggeman and Leys (2009) published another review paper on nonthermal plasma in contact with water (Bruggeman & Leys, 2009). They discussed three different types of plasmas: direct liquid discharges, discharges in gas phase with a liquid electrode, and discharges in bubbles in liquids. A different excitation method for each type was discussed individually. In addition, plasma characteristics of the different types of plasma in liquids were discussed. Currently, several research groups around the world actively study plasma discharges for water treatment, which is briefly discussed next.

1.5.2 Application of Plasma Discharges in Water

Schoenbach and his colleagues at Old Dominion University, Virginia, have studied the electrical breakdown in water with submillimeter gaps between pin and plane electrodes by using optical and electrical diagnostics with a temporal resolution on the order of 1 ns (Joshi et al., 2009; Joshi et al., 2004; Qian, Joshi, and Schoenbach, 2006; Qian et al., 2005; Schoenbach et al., 2008; Schoenbach et al., 2007). By using a Mach-Zehnder interferometer, the electric field distribution in the prebreakdown phase was determined by means of the Kerr effect, which indicates a change in the refractive index of a material. Values of electric fields in excess of computed electric fields, which reached over 4 MV/cm for applied electrical pulses of 20 ns, were recorded at the tip of the pin electrode. The results of this research found bioelectric applications in the construction of compact pulsed power generators.

Locke and his colleagues at Florida State University have qualitatively studied the production of reductive species by pulsed plasma discharge in water using different chemical probes (Locke, Burlica, & Kirkpatrick, 2006; Mededovic, Finney, & Locke, 2008; Sahni & Locke, 2006). They showed that the formation of primary radicals from water decomposition occurred in the discharge zone. The immediate region surrounding the discharge zone was responsible for radical recombination to form products that diffused into bulk water, where the radicals participated in bulk-phase reactions. The rate of the formation of reductive species in the pulsed streamer discharge increased as the input power to the system increased, offering a possibility that in a mixture of aqueous contaminants some pollutants or a component of certain pollutants could degrade by reductive mechanisms, thereby increasing the degradation efficiency of the process.

Graves and his colleagues at the University of California, Berkeley, presented a unique method to inactivate microorganisms in 0.9% NaCl solution (i.e., normal saline solution) by means of microplasmas (Sakiyama et al., 2009). They employed *E. coli* bacteria to investigate the disinfection efficiency of the device. The device consisted of a thin titanium wire covered by a glass tube for insulation except for the tip of the wire and ground electrode. Microbubbles were formed at both electrodes from the application of an asymmetric high-frequency high voltage. Repetitive light emission was observed in the vicinity of the powered electrode. More than 99.5% of *E. coli* was deactivated in 180 s.

Sato and his colleagues at Gunma University, Japan, studied the environmental and biotechnological applications of high-voltage pulsed discharges in water (Sato and Yasuoka, 2008; Sato, Yasuoka, & Ishii, 2008, 2010; Sato, 2008; Sato, Tokutake, et al., 2008). A pulsed discharge was formed in water by applying a high-voltage pulse in point-to-plane electrode systems. They found that bubbling through a hollow needle electrode made it possible to raise the energy efficiency in the decomposition of organic materials by reducing the initial voltage of the discharge. The bubbling of oxygen gas was effective for the decomposition because of the formation of active species originating from oxygen gas.

Sunka and other researchers from the Institute of Plasma Physics, Academy of Sciences of the Czech Republic, developed a pulsed corona discharge generator in water using porous ceramic-coated rod electrodes (Lukes et al., 2008, 2009; Sunka et al., 2004). They studied the properties of the ceramic layer and its interaction with the electrolyte and reported that surface chemistry at the electrolyte/ceramic surface interface was an important factor in generating electrical discharges in water using porous ceramic-coated electrodes. Initiation of the discharge in water using these types of electrodes depended on the surface charge of the ceramic layer in addition to the permittivity and porosity of the ceramic layer. The surface charge could be determined by the polarity of applied voltage and the pH and the chemical composition of aqueous solution. By applying bipolar high-voltage pulses to eliminate possible buildup of an electrical charge on the ceramic surface, a large-volume plasma could be produced in water in the range of kilowatts.

2

Generation of Plasma in Liquid

2.1 Introduction

As briefly mentioned in the first chapter, when one considers the mechanism of plasma discharge in water, there can be two different approaches: The first approach divides the breakdown to a bubble process and an electronic process (Akiyama, 2000), while the second approach classifies the electric breakdown in water into partial discharge and a full discharge, such as arc or spark. In this chapter, we focus on the second approach; the first approach is discussed in Chapter 3.

2.2 Partial and Full Discharges in Liquid

Electrical discharges in liquid are usually divided into partial and full discharges (Sato, Ohgiyama, and Clements, 1996; Sugiarto, Ohshima, and Sato, 2002; Sun, Sato, and Clements, 1999; Sugiarto et al., 2003; Akiyama, 2000; Lisitsyn et al., 1999b; Katsuki et al., 2002; Manolache et al., 2001; Manolache, Shamamian, and Denes, 2004; Ching et al., 2001; Ching, Colussi, and Hoffmann, 2003; Destaillats et al., 2001). If the discharge does not reach the second electrode, it is called a partial discharge (also called pulsed corona discharge, in analogy with the discharges in gases), and branches of such a discharge are called streamers. The nature of the discharges in liquids and the mechanism of streamer formation are much less understood and may be completely different from those for discharges in gases. If a streamer reaches the opposite electrode, it makes a conductive channel between the two electrodes and is usually called a full discharge. Furthermore, if the current through the discharge is very high (above 1 kA), it is called an arc discharge. While an arc discharge is usually continuous, the transient phase of the arc discharge is referred to as a pulsed spark discharge. In the partial discharge, the current is transferred by slow ions, producing corona-like discharges (i.e., nonthermal plasma). For a case of a liquid with a high electrical conductivity, a larger discharge current flows, resulting in shortening of the streamer length due to the faster compensation of the space charge electric fields on the head of the streamer. Subsequently, a higher power density (i.e., a higher plasma density) in the channel is obtained, resulting in a higher plasma temperature, higher ultraviolet (UV) radiation, and the generation of acoustic waves. In the full discharges, such as arc or spark, the current is transferred by electrons. The high current heats a small volume of plasma in the gap between the two electrodes, generating a quasithermal plasma, with the temperatures of electrons and heavy particles almost equal.

To generate partial or full electrical discharges in liquid, usually one needs to have a pulsed high-voltage (HV) power supply with a voltage rise time shorter than the Maxwellian relaxation time of the liquid. High electric field strength can usually be achieved by using needle electrodes with sharp tips, from which electric discharges in water usually start. Two simple geometries are shown in Figure 2.1: point-to-plane geometry and point-to-point geometry. The characteristics of pulsed corona and pulsed arc (Figure 2.2) are summarized in Table 2.1.

2.2.1 Thermal Breakdown Mechanism

When a voltage pulse is applied to water, it induces a current and redistribution of the electric field. More specifically, the voltage pulse, once applied, immediately stimulates the rearrangement of electric charges in water, and the rearrangement of the electric charges results in fast redistribution of electric field in water. Due to the dielectric nature of water, an electric double layer is formed near the electrode, resulting in the localization of the major portion of the applied electric field in the vicinity of the electrode. This electric field can become high enough for the formation of a narrow conductive channel, which is heated up by electric current to temperatures of about 10,000 K. Thermal plasma generated in the channel is rapidly expanding

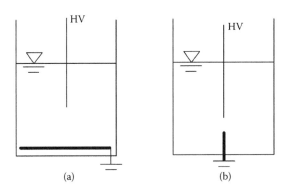

(a)   (b)

FIGURE 2.1
Two common electrode configurations for plasma discharge in water: (a) point to plane; (b) point to point.

(a) (b)

FIGURE 2.2
(See color insert.) Images of plasma discharge in water: (a) pulsed corona; (b) pulsed arc.

from the narrow channel into water, forming a plasma bubble. High electric conductivity in the plasma channel leads to the shifting of the high electric fields from the channel to the bubble. These electric fields provide a drift of negatively charged particles from the bubble into the channel. Taking into account that the temperature in the plasma bubble is not large enough to cause thermal ionization, and the electric field at the bubble is not sufficient to cause direct electric impact ionization, the oxygen-containing negative ions from water are believed to make major contributions in the negative charge transfer from the bubble into the channel.

The plasma bubble can be characterized by both a very high temperature gradient and a large electric field. The energy required to form and sustain the plasma bubble is provided by joule heating in the narrow conductive channel in water. High current density in the channel is limited by the conductivity in the relatively cold plasma bubble, where temperature is about 2,000 K. The electric conductivity in the bubble is determined not by electrons but by negative oxygen-containing ions. Further expansion of the plasma bubble leads to the cooling of the bubble and decrease of the charged particle densities. Subsequently, the electric current decreases, resulting in a significant reduction in joule heating in the conductive channel in water and eventual cooling of the channel itself.

The physical nature of thermal breakdown can be related to thermal instability of local leakage currents through water with respect to the joule overheating. If the leakage current is slightly higher at one point, the joule heating and, hence, temperature increase there. The increasing temperature results in a significant growth of local electric conductivity and leakage current. Temperature increases exponentially to several thousand degrees at a local point, leading to the formation of the narrow plasma channel in water, which results in the thermal breakdown. The thermal breakdown is a critical thermoelectric phenomenon taking place at applied voltages exceeding a certain threshold value, when heat release in the conductive channel cannot be compensated by heat

TABLE 2.1

Summary of the Characteristics of Pulsed Corona, Pulsed Arc, and Pulsed Spark Discharge in Water

---

1. Pulsed corona
   - Streamer channels do not propagate across the entire electrode gap (i.e., partial electrical discharge)
   - Streamer length: order of centimeters.
   - Streamer channel width: 10–20 μm.
   - The current is transferred by ions.
   - Nonthermal plasma.
   - Weak-to-moderate UV generation.
   - Relatively weak shock waves.
   - Treatment area is limited at a narrow region near the corona discharge.
   - A few joules per pulse, often less than 1 J per pulse.
   - Operating frequency is in a range of 100–1,000 Hz.
   - Relatively low current, that is, peak current is less than 100 A.
   - Electric field intensity at the tip of the electrode is 100–10,000 kV/cm.
   - A fast-rising voltage.
2. Pulsed arc
   - The current is transferred by electrons.
   - Quasithermal plasma.
   - An arc channel generates strong shockwaves within cavitation zone.
   - High current filament channel bridges electrode gap.
   - Channel propagates across entire electrode gap.
   - Gas inside channel is ionized.
   - Strong but short-lived UV emission and high radical density.
   - A smaller gap between two electrodes of about 5 mm is needed compared to that in pulsed corona.
   - Light pulse from spark discharge includes about 200-nm wavelength.
   - Time delay between voltage pulse increase and spark formation depends on both capacitance size and the electric conductivity of water.
   - Large energy discharges, greater than 1 kJ per pulse, desired for wastewater treatment.
   - Large current, on the order of 100 A, with a peak current greater than 1,000 A.
   - Voltage rise time is in a range of 1–10 μs.
   - Pulse duration about 10 ms.
   - Temperature of the arc is greater than 10,000 K.
3. Pulsed spark
   - Similar to pulsed arc, except for a shorter pulse duration and lower temperature.
   - Pulsed spark is faster than pulsed arc (i.e., strong shockwaves are produced).
   - Plasma temperature in the spark is around a few thousand kelvin.

---

transfer losses to the surroundings. The described sequence of plasma channel events takes place in frameworks of a single voltage pulse. When the next voltage pulse is applied to water, a new thermal breakdown and new microarc occur in another location on the surface of the electrode.

During the plasma discharge, the thermal condition of the water is constant. For water relatively far away from the discharge, it stays in a liquid state with a thermal conductivity of about 0.68 W/mK. When joule heating between the two electrodes is larger than a threshold value, an instability can occur, resulting in instant evaporation and a subsequent thermal breakdown.

On the other hand, when joule heating is smaller than the threshold value, nothing happens but electrolysis; hence, the breakdown never takes place. Since the joule heating is inversely proportional to the resistance of matter when a fixed voltage is applied between the two electrodes, the resistance is inversely proportional to the electric conductivity of the dielectric medium (here initially liquid water and later water vapor).

To analyze the thermal instability, it can be assumed that electric conductivity of water σ_e can be expressed as an exponential function of temperature T:

$$\sigma_e = \sigma_0 e^{-(E_a/RT)} \tag{2.1}$$

where E_a is an activation energy, σ_0 is the initial electric conductivity, and R is the universal gas constant. When the temperature of the medium increases, the electric conductivity of dielectric medium increases, resulting in the decrease in the resistance. Thus, the joule heating increases, increasing the temperature of the dielectric medium. Subsequently, the increased temperature increases the electric conductivity, further increasing temperature, leading to a thermal "explosion" that can be referred to as an instability and described by linear perturbation analysis of the transient energy conservation equation:

$$\rho C_p \frac{\partial T}{\partial t} = \sigma_0 e^{-\left(\frac{E_a}{RT}\right)} \left(E^2 - \lambda \nabla^2 T\right) \tag{2.2}$$

where rC_p is the thermal mass per unit volume, E (V/cm) is the electric field, and l is the thermal conductivity of water. The second term in the right-hand side represents heat conduction, which takes place with a large temperature gradient along the radial direction. The minus sign in the second term means that it represents heat loss to the surrounding water. Note that the convection heat loss is not considered because there is no time for heat to dissipate via convection.

The instability is usually described in terms of its increment W, which is an angular frequency (rad/s). When Ω is greater than zero, the perturbed temperature exponentially increases with time, resulting in thermal explosion. When Ω is less than zero, the perturbed temperature exponentially decreases with time, resulting in a steady-state condition. When Ω is complex, the perturbed temperature oscillates with time. The linear perturbation analysis of Equation 2.2 leads to the following expression for the increment of the thermal breakdown instability (Fridman, Gutsol, & Cho, 2007):

$$\Omega = \left[\frac{\sigma_0 E^2}{\rho C_p T_0}\right] \frac{E_a}{RT_0} - D\frac{1}{R_0^2} \tag{2.3}$$

where R_0 is the radius of the breakdown channel, $D \approx 1.5 \times 10^{-7}$ m²/s is the thermal diffusivity of water, $C_p = 4{,}179$ J/kg·K, and l is 0.6 W/m·K. The first

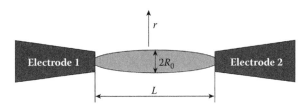

FIGURE 2.3
Sketch of a plasma channel between two electrodes surrounded by water.

term in the right-hand side is made up of the product of two, $[\sigma_0 E^2/\rho C_p T_0]$ represents the frequency of heating as the numerator is joule heating, whereas the denominator is the heat stored in the water medium; $[E_a/RT_0]$ represents the ratio of the activation energy to temperature, a sensitivity indicator. The second term in the right-hand side represents the ratio of the thermal diffusivity to the square of the characteristic length for radial heat conduction (see Figure 2.3), indicating how fast heat dissipates along the radial direction. The first term is only active during the period when the pulse power is on, while the second term is always active, even during the period when the pulse power is turned off. Hence, there is a balance between the joule heat generation by pulse discharges and heat conduction to the surrounding water. When the heat generation is greater than the conduction loss, the increment Ω becomes positive, leading to the thermal explosion. Hence, the critical phenomenon leading to the thermal explosion is given as follows:

$$\left[\frac{\sigma_0 E^2}{\rho C_p T_0}\right]\frac{E_a}{RT_0} \geq D\frac{1}{R_0^2} \tag{2.4}$$

Note that $\Omega = 0$ means the transition from the stabilization to thermal explosion, a condition that can be defined as the critical phenomenon.

Since the electric conductivity s of a dielectric medium is extremely sensitive to temperature, as shown in Equation 2.1, one can expect that as the temperature increases, the breakdown voltage would decrease.

The breakdown voltage V is given by the product of electric field strength E and the distance between two electrodes L. Thus, one can rewrite Equation 2.4 as

$$\left[\frac{\sigma_0 (EL)^2}{\rho CpT_0}\right]\frac{E_a}{RT_0} = \left[\frac{\sigma_0 V^2}{\rho C_p T_0}\right]\frac{E_a}{RT_0} \geq D\frac{1}{R_0^2/L^2} \tag{2.5}$$

If we introduce a geometry factor, $G = L/R_0$, one can rewrite the above equation as

$$\left[\frac{\sigma_0 V^2}{\rho C_p T_0}\right]\frac{E_a}{RT_0} \geq DG^2 \tag{2.6}$$

From this equation, the breakdown voltage V can be obtained as

$$V \geq \sqrt{\frac{kRT_0^2}{\sigma_0 E_a}} G \tag{2.7}$$

For the plasma discharge in water, the breakdown voltage can be numerically estimated as follows:

$$V \geq \sqrt{\frac{kRT_0^2}{\sigma_0 E_a}} G = \sqrt{\frac{0.613 \times 461.5 \times 300^2}{0.05 \times 700}} G \approx 26G \tag{2.8}$$

In a case of $L = 1$ cm, the diameter of the streamer is usually on the order of 10 μm, leading to the geometry factor $G = L/R_0 = 1,000$, and the breakdown voltage in water becomes 26,000 V.

2.2.2 Production of Reactive Species, UV, and Shock Wave by Electrical Discharges in Liquid

When a plasma discharge is initiated between two electrodes, the medium between the two electrodes is ionized, creating a plasma channel. The plasma discharge generates UV radiation and converts surrounding water molecules into active radical species due to the high energy level produced by the discharge. The microorganisms could be effectively inactivated, while the organic contaminants could be oxidized through the contact with active radicals. The chemical kinetics of these reactions remains an area of significant research. Various active species can be considered the by-products of plasma discharge in water. The production of these species by plasma discharge is affected by a number of parameters, such as applied voltage, rise time, pulse duration, total energy, polarity, the electric conductivity of water, and so on. Among the active species, hydroxyl radical, atomic oxygen, ozone, and hydrogen peroxide are the most important ones for the sterilization and removal of unwanted organic compounds in water. Table 2.2 summarizes the oxidation potentials of various active species produced by plasma in water, which ranges from 1.78 V (hydrogen peroxide) to 2.8 V (hydroxyl radical). Note that fluorine has the highest oxidation potential of 3.03 V, whereas chlorine, which is one of the most commonly used chemicals for water decontamination, has an oxidation potential of only 1.36 V.

A detailed schematic of the chemistry relative to plasma inside or over water can be found in the review by Bruggeman and Leys (2009). Major reactions for radical production are summarized in Table 2.3.

In addition to the active species, the electrical breakdown in water produces UV radiation (both vacuum UV [VUV] and UV). Vacuum UV, as the name indicates, can only propagate in vacuum because it is strongly absorbed by air or water due to its high energy. For pulsed arc discharge, the high-temperature plasma channel can function as a blackbody radiation

TABLE 2.2

Oxidation Potential of Common Oxidative Agents and Active
Species Produced by Plasma in Water under Standard Conditions

Chemical	Reaction	Oxidation Potential (V)
O_2	$1/2O_2 + 2H^+ + 2e \rightarrow H_2O$	1.23
Cl_2	$1/2Cl_2 + e \rightarrow Cl^-$	1.36
Fe^{3+}	$Fe^{3+} + e \rightarrow Fe^{2+}$	0.77
F_2	$1/2F_2 + e \rightarrow F^-$	3.03
MnO_4^-	$MnO_4^- + 8H^+ + 5e \rightarrow Mn^{2+} + 4H_2O$	1.51
O_3	$O_3 + 2H^+ + 2e \rightarrow O_2 + H_2O$	2.07
O	$O + 2H^+ + 2e \rightarrow H_2O$	2.42
OH	$OH + H^+ + e \rightarrow H_2O$	2.81
H_2O_2	$1/2H_2O_2 + H^+ + e \rightarrow H_2O$	1.78
HO_2	$HO_2 + H^+ + e \rightarrow H_2O_2$	1.50
O_2^-	$O_2^- + H^+ \rightarrow HO_2$	1.00

TABLE 2.3

Major Reactions for Production of Reactive Species in Liquid Plasma
($T_e \approx 1$ eV, $T_g \approx 300$ K)

Reaction	Reaction Rate (m^3s^{-1})	Reference
$e + H_2O \rightarrow OH + H + e$	2.3×10^{-18} to 1.8×10^{-16}	Itikawa and Mason, 2005
$e + H_2O \rightarrow OH + H^-$	4.9×10^{-18} to 4.7×10^{-17}	Itikawa and Mason, 2005
$e + H_3O^+ \rightarrow OH + H_2$	10^{-13}	Millar, Farquhar, and Willacy, 1997
$e + H_2O^+ \rightarrow OH + H$	2.6×10^{-14}	Jensen et al., 1999
$OH + OH \rightarrow H_2O + O$	1.9×10^{-18}	Herron and Green, 2001
$OH + OH \rightarrow H_2O_2$	2.6×10^{-17}	Herron and Green, 2001
$H_2O + H_2O^+ \rightarrow OH + H_3O^+$	1.9×10^{-15}	Gordillo-Vázquez, 2008
$H^- + H_2O^+ \rightarrow OH + H_2$	2.0×10^{-13}	Kushner, 1999
$H^- + H_3O^+ \rightarrow OH + H_2 + H$	2.3×10^{-13}	Millar, Farquhar, and Willacy, 1997
$OH + O \rightarrow O_2 + H$	3.3×10^{-17}	Herron and Green, 2001
$OH + H \rightarrow H_2O$	2×10^{-16}	Herron and Green, 2001
$OH + H_2O_2 \rightarrow HO_2 + H_2O$	1.7×10^{-19}	Herron and Green, 2001

source. The maximum emittance is in the ultraviolet A (UVA) to ultraviolet
C (UVC) range of the spectrum (200–400 nm) (Robinson, Ham, and Balaster,
1973; Sunka, 2001), as determined by the Stephen-Boltzmann law. Water is
relatively transparent to UV radiation in this wavelength range as long as
the water remains clear without too many suspended particles. The energy
per photon range from 3.1 to 6.2 eV. UV radiation (Table 2.4) has proven to
be effective for decontamination processes and is gaining popularity as a
means for sterilization because chlorination leaves undesirable by-products

TABLE 2.4

Types of Ultraviolet Radiation

Name	Abbreviation	Wavelength	Energy (eV)
Ultraviolet A	UVA	315–400 nm	3.10–3.94 eV
Ultraviolet B	UVB	280–315 nm	3.94–4.43 eV
Ultraviolet C	UVC	200–280 nm	4.43–6.20
Vacuum ultraviolet	VUV	100–200 nm	6.20–12.4 eV

in water. The radiation in the wavelength range of 240–280 nm may cause irreparable damage to the nucleic acid of microorganisms, preventing proper cellular reproduction and thus effectively inactivating the microorganisms.

Alternatively, the photons can provide the necessary energy to ionize or dissociate water molecules, generating active chemical species. It has been suggested that the UV system may produce charged particles in water such that charge accumulation occurs on the outer surface of the membrane of a bacterial cell (Laroussi et al., 2002). Subsequently, the electrostatic force on the membrane overcomes the tensile strength of the cell membrane, causing its rupture at a point of small local curvature as the electrostatic force is inversely proportional to the local radius squared.

Next, the ability for the discharge to generate shock waves is briefly summarized. When an HV, high-current discharge takes place between two electrodes submerged in water, a large part of the energy is consumed on the formation of a thermal plasma channel. The expansion of the channel against the surrounding water generates a shock wave. For the corona discharge in water, the shock waves are often weak or moderate, whereas for the pulsed arc the shock waves are strong. The difference arises from the fact that the energy input in the arc or spark discharge is much higher than that in the corona.

Similarly, the arc produces much greater shock waves due to its higher energy input than spark. The water surrounding the electrodes becomes rapidly heated, producing bubbles, which help the formation of a plasma channel between the two electrodes. The plasma channel may reach a very high temperature of 14,000–50,000 K, consisting of a highly ionized, high-pressure, and high-temperature gas. Thus, once formed, the plasma channel tends to expand. The energy stored in the plasma channel is dissipated via both radiation and conduction to the surrounding cool liquid water as well as via mechanical work. At the liquid-gas phase boundary, the high-pressure buildup in the plasma is transmitted into the water interface, and an intense compression wave (i.e., shock wave) is formed, traveling at a much greater speed than the speed of sound. Note that the shock waves have another benefit in the sterilization process through a good mixing of water to be treated, significantly enhancing the plasma treatment efficiency.

However, the plasma discharge for water treatment is not without deficiencies. One of the concerns in the use of a sharp needle as an HV electrode is the adverse effect associated with the needle tip erosion (Matsushima et al., 2006). In a point-to-plane geometry, a large electric field can be achieved due to the sharp tip of the needle with a minimum applied voltage V. For a sharp parabolic tip of the needle electrode, the theoretical electric field at the needle tip becomes

$$E \propto V/r \tag{2.9}$$

where r is the radius of curvature of the needle tip. As indicated by Equation 2.9, the electric field at the tip of the electrode is inversely proportional to the radius of curvature of the needle tip. Hence, the maximum electric field could be obtained by simply reducing the radius of curvature r, which is much easier than increasing the voltage as the maximum value of the voltage is usually restricted by the electric circuit as well as insulation materials used around electrodes.

Sunka et al. (1999) pointed out that the very sharp tip anode would be quickly eroded by the discharge, and one had to find some compromise between the optimum sharp anode construction and its lifetime for extended operation. Also, it was demonstrated that the erosion of electrodes at pulse electric discharge in water would result in the production of metal and oxide nanoparticles in water (Kolikov et al., 2005, 2007). These particles are difficult to remove once they enter the drinking water system due to their nanometer sizes, and the potential danger to the human body is not clearly known.

Another concern in the application of pulsed electric discharges in water is the limitation posed by the electrical conductivity of water on the production of such discharges (Sunka et al., 1999). In the case of a low electric conductivity, below 10 μS/cm, the range of the applied voltage that can produce a corona discharge without sparking is narrow. On the other hand, in the case of a high electric conductivity, above 400 μS/cm, which is the typical conductivity of tap water, streamers become short, and the efficiency of radical production decreases. In general, the production of hydroxyl radicals and atomic oxygen is more efficient at water conductivity below 100 μS/cm. Thus, this is one of the major challenges in the application of plasma discharges for cooling water management as the electric conductivity of most cooling water is at the range of 2,000–4,000 μS/cm. Even bigger challenges exist for the treatment of seawater (for example, ballast water), for which the electric conductivity can be higher than 30,000 μS/cm.

2.3 Underwater Plasma Sources

2.3.1 Direct Discharges in Liquid

Various electrode geometries have been studied for the generation of plasma discharges in liquid. Figure 2.4 shows some of the typical electrode configurations. Note that only when both the HV electrode and ground electrode are placed in liquid are shown. In general, electrodes with a small radius of curvature r_c at the tips are used to ignite a plasma discharge because it can produce a high initiation electric field in liquid. Among them, the point-to-plane geometry has been the most commonly used configuration (shown in Figures 2.1a and 2.4a). Also, a point-to-plane geometry with multiple points was used to generate a large-volume corona discharge in water (Figure 2.4b). For pulsed arc discharges, a point-to-point electrode geometry was often used (Figures 2.1b and 2.4c). The electric field at the tip of an electrode can be estimated as U/r_c, where U is the applied voltage, and r_c is the radius of curvature at the tip. For a 100-mm diameter wire and typical 30-kV voltage for usual point configurations, the enhanced electrical field using a small r_c can reach the megavolts per centimeter required for liquid discharge initiation.

As mentioned, one of the concerns in the use of a sharp needle as the HV electrode is the tip erosion due to the intense local heating at the tip.

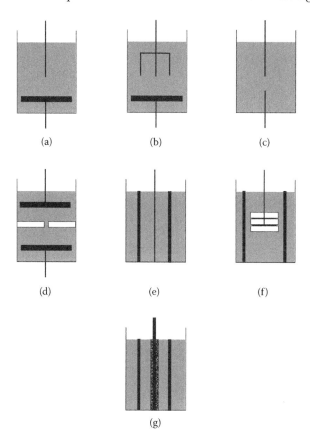

FIGURE 2.4
Schematics of electrode geometries used for plasma discharges in liquid: (a) single point to plane; (b) multiple points to plane; (c) point to point; (d) pinhole; (e) wire to cylinder; (f) disk electrode; (g) composite electrode with porous ceramic layer.

To overcome the limitation of the needle-plate configuration, other electrode systems are developed to achieve the goal of local electric field enhancement. One example is "pinhole" electrodes (also called a diaphragm discharge, as shown in Figure 2.4d), where the HV and ground electrodes are separated by a dielectric disk with a small hole at the center of the disk (Sunka et al., 2003; Krcma, 2006; Yong et al., 2010). The diameter of the pinhole typically varies from 0.1 to 1 mm. When HV is applied on the electrodes, an intense electric field could be formed around the pinhole. Subsequently, a predischarge current could be concentrated in the small hole, leading to strong thermal effects, resulting in the formation of bubbles and breakdown. Pulsed corona discharge occurs inside the bubbles at the pinhole because of the high electric field. The length of the streamers generated is decided by such parameters as water conductivity, the size of the pinhole, flow velocity through the pinhole, and voltage polarity. It was observed that when the flow rate through the pinhole was increased, the length of the streamers decreased, possibly due to the flow washing out tiny bubbles or ions from the pinhole (Yamada et al., 1998; Baerdemaeker et al., 2005; Baerdemaeker et al., 2007a; Baerdemaeker et al., 2007b). Similar to the corona discharge in the point-to-plane geometry, a pulsed arc discharge could be formed once the streamer bridges the two electrodes. Figure 2.5 shows (a) pulsed corona and (b) arc discharges through a pinhole developed at Drexel Plasma Institute. A similar design with multiple pinholes was attempted. However, it was observed that it was difficult to discharge simultaneously at each pinhole, and very high overvoltage with short rise time is usually required.

Another critical issue that researchers are facing is the scaling up of the plasma source to increase the volume of an active plasma discharge region for industrial applications with a high water flow rate. Clearly, the point-to-plane electrode geometry would be difficult to scale up for such industrial applications. Also, it is difficult to discharge uniformly at multiple pinholes. To treat a large volume of water with plasma discharges effectively, different

(a)   (b)

FIGURE 2.5
(See color insert.) Images of plasma discharges through a pinhole: (a) pulsed corona; (b) pulsed arc produced at Drexel Plasma Institute.

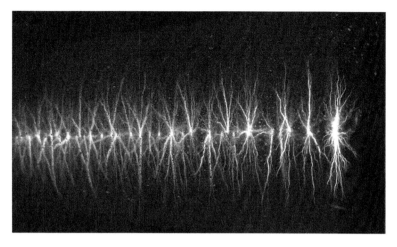

FIGURE 2.6
Time-integrated image of discharges generated using a wire-cylinder geometry in water; tungsten wire and stainless steel mesh cylinder were used. Chamber dimensions: 44 mm inside diameter, 100 mm long. (From Malik, M.A., Minamitani, Y., Xiao, S., Kolb, J.F., and Schoenbach, K.H. (2005) Streamers in water filled wire-cylinder and packed bed reactors. *IEEE Trans. Plasma Sci.* 33, 490–491.)

approaches could be used, including a wire-cylinder geometry (Figures 2.4e and 2.6), a disk geometry (Figures 2.4f and 2.7), and a concentric cylinder geometry with an HV center composite electrode coated by a thin layer of porous ceramic (Figures 2.4g and 2.8).

Figure 2.6 shows the corona discharge generated using a wire-cylinder-type electrode system developed by Malik et al. (2005). A thin tungsten wire 0.075 mm in diameter was fixed along the axis of a stainless steel tube. A positive HV pulse of up to 90 kV was applied to the wire using a Marx generator. A pulse duration of 500 ns full-width half-maximum (FWHM) with a rise time of 80–150 ns and a 400-ns fall time was used so that the streamers emerging from the wire electrode did not reach the outer electrode and consequently could not generate an arc. The energy deposited in the coaxial reactor was 11 J.

The geometry using multiple disks shown in Figure 2.4f utilized a number of thin, circular stainless steel disk electrodes separated by dielectric layers to produce pulsed multichannel discharges in water. The thickness of the disk electrodes was about 20 μm. An outer cylindrical stainless steel case was used as the ground electrode, and the gap distance between the inner wall of the cylinder and the edge of the acrylic disks was 5 mm.

The electric field at the edge of the stainless steel disk can be estimated as $E \sim 2U/d$, where U is the applied voltage, and d is the disk thickness. Hence, a high electric field on the order of 10^6 to 10^7 V/cm could be easily achieved, which is comparable to the electric field at the tip of a point electrode. The electric field strength would stay relatively constant throughout the discharge process as long as the thickness of the disk stays constant, thus eliminating the concern for the decay of the electric field due to the erosion of a point electrode.

The stainless steel disks were sandwiched between pairs of acrylic disks with a diameter of 105 mm and a thickness of 5 mm. The diameter of the acrylic disk was slightly greater than that of the stainless steel disk, so when HV was applied on the stainless steel disk, the ionic prebreakdown current was limited to a small area enclosed by the acrylic disks and the edge of the stainless steel disk. Limiting the current to the small area allowed water to be heated and vaporized through joule heating, thus effectively promoting the initiation of the plasma discharges in these vaporized areas.

The mechanism is similar to that of the pinhole discharge diaphragm mentioned. However, the diaphragm discharge is usually produced through a small hole, leading to a limited treatment capacity. The current design using thin stainless steel disks sandwiched between acrylic disks allows the generation of plasma along the periphery of the disk, resulting in a much larger treatment volume. Furthermore, the entire electrode system can be easily scaled up by stacking multiple metal disks for water treatment at a high flow rate. Figure 2.7 shows photographs of pulsed multichannel discharge arrays generated with one (single-layer) and two (double-layer) stainless steel disks.

Sunka and his coworkers developed an HV composite electrode coated by a thin layer of porous ceramic to produce a large-volume corona discharge in water (Lukes et al., 2008). Such an electrode can be used in a wide variety of geometrical configurations, including cylindrical wire and planar geometry. The role of the ceramic layer is to enhance the electric field on the anode surface by the concentration of the predischarge current in small open pores so that a large number of discharge channels can be distributed uniformly and homogeneously on the electrode surface. The composite electrodes can be made in various dimensions, enabling the construction of reactors that can operate at an average power on the order of kilowatts. Figure 2.8 shows images of multichannel pulsed

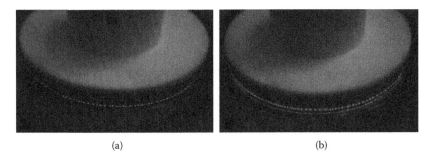

(a)   (b)

FIGURE 2.7
Pulsed multichannel discharge array in water generated by two stainless steel disk electrodes separated by dielectric layer: (a) single layer; (b) double layers. (Yang, Kim et al., 2011a.)

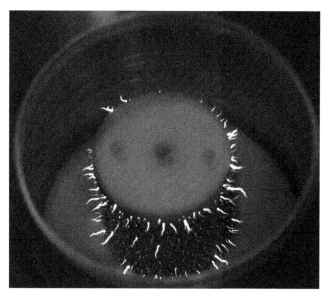

FIGURE 2.8
Multichannel pulsed electrical discharge in water generated using porous-ceramic-coated metallic electrodes. (From Lukes, P., Clupek, M., Babicky, V., and Sunka, P. (2009) The role of surface chemistry at ceramic/electrolyte interfaces in the generation of pulsed corona discharges in water using porous ceramic-coated rod electrodes. *Plasma Processes Polym.* 6, 719–728.)

electrical discharges in water generated using porous-ceramic-coated metallic electrodes.

2.3.2 Bubble Discharges in Liquid

In engineering applications of plasma discharges in liquids, HV, high-power discharges are often needed for the generation of breakdown in liquids as well as for desired processing. In these cases, the high energy supplied by a power source is first used to evaporate the liquid adjacent to the HV electrode, generating gas bubbles that are subsequently ionized by large electric fields produced by the HV. Liquid temperatures in such applications are usually high, at least locally, near the breakdown locations due to the excessive power dissipated in the liquid. However, in some circumstances high temperature is not desired. For such applications, a nonthermal plasma system that can generate gas-phase plasmas in contact with liquids is often desired. From a practical point of view, these discharges hold another advantage because gas-phase plasmas are much easier to produce than direct plasma in liquid. Since the gas-phase plasma can only interact with the liquid through the gas-liquid interface, a maximization of the interface area is usually desired, which can be achieved using bubble plasmas (i.e., plasmas generated in small bubbles suspended in liquid). Note that the ratio of the area of

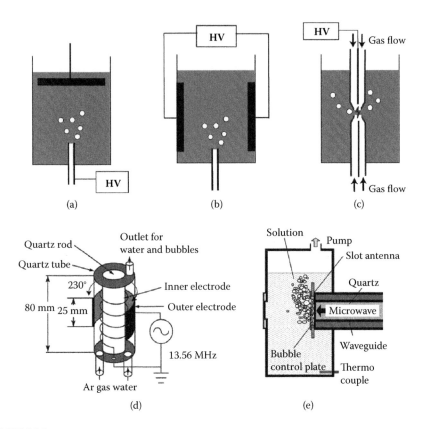

FIGURE 2.9
Schematics of electrode geometries for bubble discharge: (a) point to plane; (b) parallel plate; (c) gas channel with liquid wall; (d) RF bubble discharge. (From Hironori, A. (2008) Plasma generation inside externally supplied Ar bubbles in water. *Plasma Sources Sci. Technol.* 17(2), 025006.) (e) Microwave bubble discharge. (Ishijima, T. (2010) Efficient production of microwave bubble plasma in water for plasma processing in liquid. *Plasma Sources Sci. Technol.* 19(1), 015010.)

the gas-liquid interface to the total gas volume is inversely proportional to the radius of the gas bubbles. Some of the examples are shown in Figure 2.9.

Similar to direct discharges in water, the most commonly used configuration is the point-to-plane configuration, where the point electrode is made of a small-diameter hollow tube to inject gas into water. Different types of gas were used depending on applications. For example, oxygen gas was often used to promote the formation of oxygen radicals.

Alternatively, gas was bubbled between two metal electrodes (Figure 2.9b). The discharge occurred between the electrodes by applying the HV, producing OH radicals that were detected by a spectroscopic technique (Miichi et al., 2000).

Another interesting discharge in liquid was to use a gas channel, inside of which two metal electrodes were placed to generate plasma discharge

(Figure 2.9c). The gas was continuously supplied through the hollow tube, flowing around the electrodes from both sides and exiting from the open ends at the middle of the reactor (see Figure 2.9c). The gases coming from the top and bottom merged into one where two point electrodes were closely positioned, forming a stable gas channel between the two metal electrodes. Subsequently, the generated discharge was an arc discharge that was cooled and stabilized by the surrounding water.

Aoki and his coworkers studied radio-frequency (RF)-excited discharges in argon bubbles in a dielectric-covered metal rod and wire reactor (Figure 2.9d). First, bubbles were formed in front of the slot antenna (see black area in the figure) through the microwave heating of water; water in an evacuated vessel at a vapor pressure of 5 kPa was evaporated by a slight increase in the temperature above the boiling point. In the second step, microwave breakdown took place inside bubbles filled with water vapor. In the third step, the bubbles containing the plasma moved up due to the upward force by buoyancy. Finally, new water filled the space vacated by the bubbles in front of the slot antenna. These steps were successively repeated, forming a large number of bubble plasmas. Microwave-excited plasma in water with or without externally introduced bubbles (Figure 2.9e) was studied by Ishijama (2010).

3

Bubble and Electronic Initiation Mechanism

3.1 Introduction

Since the very early stage of research related to plasma in liquid, two radically different theories for the electric breakdown of liquid were proposed to describe the initiation process. According to the first approach, the electric discharge in liquid initiates from gas cavities either preexisting in the liquid or on the electrodes or newly formed through electrolysis, boiling, or electron decomposition. This theory is usually referred to as the *bubble mechanism*. According to the other theory, the discharge is a consequence of avalanche multiplication of free charge carriers in the liquid similar to the gas discharge model generalized to the liquid phase, in which electrons in strong fields are accelerated, ionizing molecules and atoms. This discharge theory is conventionally called the electronic mechanism.

In this chapter, the two competing theories are discussed. First, we start from the well-established classical plasma discharge physics in gas phase to introduce some background knowledge.

3.2 Electrical Breakdown in Gas Phase

3.2.1 The Townsend Breakdown Mechanism

Electric breakdown in the gas phase is a complicated multistage threshold process that occurs when the electric field exceeds some critical value. During the short breakdown period, typically 0.01 to 100 µs, the gas becomes conductive and, as a result, generates different kinds of plasmas. The breakdown mechanisms can be sophisticated, but all of them usually start with the electron avalanche (e.g., the multiplication of some primary electrons in cascade ionization) (Raizer, 1997).

Consider first the simplest breakdown in a plane gap with an interelectrode distance d, connected to a direct current (DC) power supply with voltage V. The two electrodes provide a quasihomogeneous electric field $E = V/d$. It is

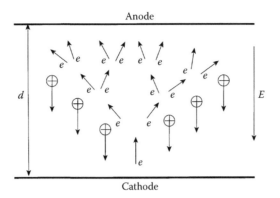

FIGURE 3.1
Townsend breakdown gap.

apparent that an occasional formation of primary electrons near the cathode may occur, providing a very low initial current i_0. Each primary electron drifts to the anode, concurrently ionizing the gas between the electrodes and generating an avalanche (see Figure 3.1). The avalanche evolves in both time and space because the multiplication of electrons proceeds along with their drift from the cathode to the anode.

It is convenient to describe the ionization in an avalanche, not by the ionization rate coefficient, but rather by the Townsend ionization coefficient α, which can be defined as the electron production per unit length or the number of ionizations per unit length. Accordingly, the rate of change of the electron density with respect to x becomes

$$\frac{dn_e}{dx} = an_e \tag{3.1}$$

Thus, the multiplication of electrons (with an initial electron density n_{e0}) per unit length along the electric field can be given as

$$n_e(x) = n_{e0} \cdot \exp(ax) \tag{3.2}$$

The Townsend ionization coefficient is related to the ionization rate coefficient $k_i(E/n_0)$ and electron drift velocity v_d as

$$\alpha = \frac{v_i}{v_d} = \frac{k_i(E/n_0)n_0}{v_d} = \frac{1}{\mu_e}\frac{k_i(E/n_0)}{E/n_0} \tag{3.3}$$

where v_i is the ionization frequency with respect to a single electron, and μ_e is the electron mobility (defined as v_d/E). Noting that breakdown begins at room temperature and that electron mobility is inversely proportional to pressure, it is convenient to present the Townsend coefficient as the similarity parameter α/p depending on the reduced electric field E/p.

According to the definition of Townsend coefficient a, each primary electron generated near the cathode produces $\exp(\alpha d) - 1$ positive ions in the gap. (Note that here the electron losses due to recombination and attachment to electron-negative molecules were neglected for simplicity.) All the positive ions produced in the gap per one electron are moving toward the cathode and altogether produce $\gamma[\exp(\alpha d) - 1]$ electrons from the cathode, a process described as the secondary electron emission. Here, another Townsend coefficient γ is the secondary emission coefficient, characterizing the probability of the secondary electron generation on the cathode by an ion impact. The secondary electron emission coefficient γ depends on cathode material, the state of the surface, the type of gas, and bombarding ion energy (which depends on reduced electric field E/p). Typical values of γ in discharge range from 0.001 to 0.1.

Taking into account the current of primary electrons i_0 and the electron current due to the secondary electron emission from the cathode, the total electronic part of the cathode current i_{cath} can be written as

$$i_{cath} = i_0 + \gamma i_{cath}\left[\exp(ad) - 1\right] \tag{3.4}$$

The total current in the external circuit is equal to the electronic current at the anode because of the absence of ion current. For this reason, the total current can be found as

$$i = i_{cath}\exp(ad) \tag{3.5}$$

Equations 3.4 and 3.5b lead us to the Townsend formula, which was first derived by J. S. Townsend in 1902 to describe the ignition of electric discharges:

$$i = \frac{i_0 \exp(ad)}{1 - \gamma\left[\exp(ad) - 1\right]} \tag{3.6}$$

The current in the gap is non-self-sustained as long as the denominator in Equation 3.6 is positive. As soon as the electric field, and therefore the Townsend coefficient α, becomes sufficiently large, the denominator in Equation 3.6 goes to zero, indicating transition to self-sustained current and thus breakdown taking place. Thus, the simplest breakdown condition in the gap can be expressed as

$$\gamma\left[\exp(ad) - 1\right] = 1 \tag{3.7}$$

or

$$ad = \ln\left(\frac{1}{\gamma} + 1\right) \tag{3.8}$$

This process of the ignition of a self-sustained current in a gap controlled by the secondary electron emission from the cathode is usually referred to as the Townsend breakdown mechanism.

TABLE 3.1

Numerical Parameters A and B for Calculation of Townsend Coefficient α

Gas	A (1/cm·Torr)	B (V/cm·Torr)	Gas	A (1/cm·Torr)	B (V/cm·Torr)
Air	15	365	N_2	10	310
CO_2	20	466	H_2O	13	290
H_2	5	130	He	3	34
Ne	4	100	Ar	12	180
Kr	17	240	Xe	26	350

Source: Fridman, A. (2008) *Plasma chemistry,* Cambridge University Press, Cambridge.

To derive relations for the breakdown voltage and electric field based on Equation 3.8, we can rewrite Equation 3.3 for the Townsend coefficient α in the following conventional semiempirical way:

$$\frac{\alpha}{p} = A exp\left(-\frac{B}{E/p}\right) \tag{3.9}$$

Equation 3.9, initially proposed by Townsend, relates the similarity parameters α/p and E/p. The parameters A and B for numerical calculations of α in different gases at E/p = 30–500 V/cm·Torr are given in Table 3.1.

Combining Equations 3.8 and 3.9 gives the following formulas for calculations of breakdown voltage and breakdown-reduced electric field as functions of the important similarity parameter pd:

$$V = \frac{B(pd)}{C + \ln(pd)} \tag{3.10}$$

$$\frac{E}{p} = \frac{B}{C + \ln(pd)} \tag{3.11}$$

In these formulas, parameter A is replaced by $C = \ln A - \ln\ln(1/\gamma + 1)$, taking into account the weak, double logarithmic influence of the secondary electron emission.

The breakdown voltage dependence on the parameter pd, as described by Equations 3.10 and 3.11, is usually referred to as the Paschen curve. The experimental Paschen curves for different gases are presented in Figure 3.2.

The curves in Figure 3.2 have minimum voltage points corresponding to the easiest breakdown conditions that can be found from Equation 3.11:

$$V_{min} = \frac{eB}{A}\ln\left(1 + \frac{1}{\gamma}\right) \tag{3.12}$$

$$\left(\frac{E}{p}\right)_{min} = B \tag{3.13}$$

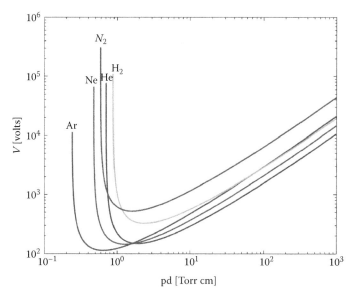

FIGURE 3.2
(See color insert.) Paschen curves for different gases. (From Fridman, A. (2008) *Plasma chemistry*, Cambridge University Press, Cambridge.)

$$(pd)_{min} = \frac{e}{A} \ln\left(1 + \frac{1}{\gamma}\right) \tag{3.14}$$

where $e = 2.72$ is the base of natural logarithm.

3.2.2 Spark Breakdown Mechanism

The Townsend mechanism described in the previous section can only be applied for quasihomogeneous breakdowns for low pressures and short discharge gaps. Another breakdown mechanism, the so-called spark breakdown mechanism or streamer breakdown mechanism, takes place in larger gaps at high pressures. In contrast to the Townsend mechanism, the spark mechanism provides breakdown in a local narrow channel with a very high current and current density, but without direct relation to the electrodes and secondary electron emission.

Both spark and Townsend breakdowns are primarily related to avalanches. However, breakdowns in large gaps cannot be considered independent and stimulated by electron emission from the cathode. The spark breakdown at high *pd* and considerable overvoltage develops much faster than the time necessary for ions to cross the gap and provides the secondary electron emission. The high conductivity spark channel can be formed even faster than electron drift time from cathode to anode. In this case, the breakdown voltage is independent of the cathode material.

The mechanism of spark breakdown is based on the concept of streamer. The spark breakdown occurs at a thin ionized channel that grows quickly along the positively charged trail left by an intensive primary avalanche between the two electrodes. The avalanche also generates photons, which in turn initiate numerous secondary avalanches in the vicinity of the primary one. Electrons of the secondary avalanches are pulled by the strong electric field into the positively charged trail of the primary avalanche, creating a streamer propagating quickly between the two electrodes. The streamer theory was originally developed by Raether (1964), Loeb (1960), and Meek and Craggs (1978).

As mentioned in the previous section, each primary electron generated near the cathode produces $\exp(\alpha x) -1$ ($\approx \exp(\alpha x)$ when distance is large) positive ions in the gap. In this case, $\exp(\alpha x)$ can be regarded as the ratio of charge amplification, and considerable space charge can be created with its own significant electric field E_a, which should be added to the external field E_0. The electrons are at the head of the avalanche, while the positive ions remain behind, creating a dipole.

The external electric field distortion due to the space charge of the dipole is depicted in Figure 3.3. In front of the avalanche head, the external E_0 and internal E' electric fields make a total field stronger, which in turn accelerates ionization. Conversely, between the separated charges, or "inside the avalanche," the total electric field is reduced, which slows the ionization. At some critical value of αx, the electric field induced by charge $N_e \approx \exp(\alpha x)$ reaches the value of the external field E_0. Then, the avalanche-to-streamer transformation takes place when the internal field of an avalanche becomes comparable with the external electric field (e.g., when the amplification parameter αx is sufficiently large).

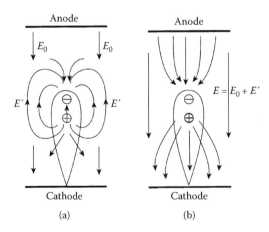

FIGURE 3.3
Electric field distribution in an avalanche.

As soon as the avalanche head reaches the anode, the electrons sink into the anode, and it is mostly the ionic trail that remains in the discharge gap. The electric field distortion due to the space charge in this case is shown in Figure 3.4. Because electrons are no longer present in the gap, the total electric field is due to the external field, the ionic trail. The field can also be viewed as a result of the ionic charge image in the anode. The resulting electric field in the ionic trail near the anode is less than the external electric field. But, at spots farther away from the electrode, it exceeds E_0, inducing the formation of so-called cathode-directed streamer.

The mechanism of the formation of a cathode-directed streamer is illustrated in Figure 3.5. High-energy photons emitted from the primary avalanche provide photoionization in its vicinity, which initiates secondary avalanches. Electrons of the secondary avalanches are pulled into the ionic trail of the primary one due to the enhanced electric field (i.e., $E_0 + E'$), as shown in Figure 3.4. The process repeats, providing the growth of the streamer like a "needle" from the anode. Obviously, the electric field at the

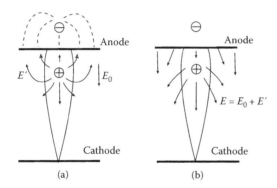

FIGURE 3.4
Electric field distribution when the avalanche reaches the anode.

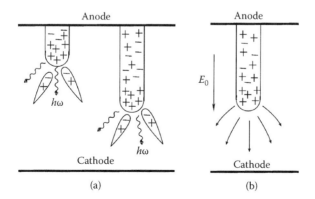

FIGURE 3.5
Cathode-directed streamer.

tip of the "anode needle" is large, which stimulates the fast streamer propagation in the direction of the cathode.

Note that streamers are not always cathode directed. If the overvoltage is high or the electrode distance is large, the avalanche-to-streamer transformation can take place before the avalanche head reaches the anode. In this case, the so-called anode-directed streamer is able to grow toward both electrodes.

In 1940, Meek proposed his criterion for streamer formation on a quantitative basis, which can be presented as the requirement for the avalanche amplification parameter αd to exceed a critical value of 20. This convenient and important criterion is known as Meek's criterion.

3.3 Electron Avalanche for Electrical Breakdown in Liquid Phase

Compared to electrical breakdown in gas, there is no adequate model of discharge development in liquid phase. While electron avalanche ionization and streamer mechanisms were generally accepted to explain the breakdown in gases, the nature of the breakdown in dielectric liquids is still disputed. The main problem is the completely different interaction dynamics in liquid systems compared to the gas-phase systems. In liquids, multiparticle interactions cannot be neglected. These interactions lead to two basic phenomena: (1) Electron-molecule interaction cross sections measured in the gas phase cannot be used in liquids; for example, cross-section shapes and even thresholds change because of molecule-to-molecule interactions. (2) The Boltzmann equation for the electron energy distribution function (EEDF) has to take into account multiparticle collisions; in other words, a collision integral cannot be calculated using a binary collision approximation in this case.

Thus, we have to note that now we have no theoretical approaches to model the discharge development in condensed liquid media. Nevertheless, there are two models usually used for qualitative analysis: "dense gas" approximation and "semiconductor" approximation. The dense gas approach assumes that the liquid is a gas with a high particle number density. This approach neglects both fundamental problems of the process description in condensed media and could lead to unpredictable errors. However, since the model can provide at least qualitative estimations for the process, we will be able to use this approach to estimate the discharge parameters. The semiconductor approach approximates liquid as a solid-state crystal with semiconductor properties. The current through the liquid media can be described on the basis of tunneling of electron-hole pairs in structured potential. This model also has obvious drawbacks but in some cases can provide a reasonable estimation of conductive current through dielectric liquids. We briefly review both approximations.

3.3.1 Dense Gas Approximation

The dominant collisions in the gas phase usually involve two particles, although under some circumstances three-body interactions can be important. In liquid, the different electron-molecule interaction behavior rises mostly because of multibody collisions due not only to the much higher liquid density but also the stronger chemical bonds between neutral particles. To make the case simpler, the dense gas approximation assumes that two-body collisions are still applicable: There is no fundamental difference between liquid and gas, and the only different parameter is the density of neutral particles.

It remains, at the present time, one of the greater challenges of atomic physics to understand the complex interactions involved when atoms or molecules are ionized by electron impact. The data on the interaction between electron and liquid molecules came mainly from e-beam or radiolysis experiments, and cross sections for the high-density liquids became known only recently (see Figure 3.6). From the collisional data, one can obtain the fluid transport

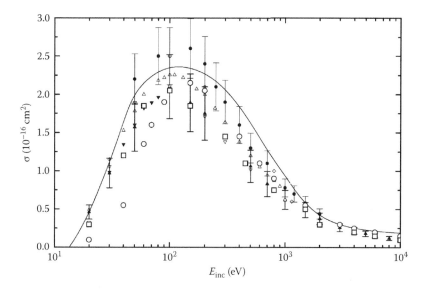

FIGURE 3.6

Comparison between the calculated ionization cross section of the vapor-water molecule by Champion (2003) (solid line) and experimental data: Bolorizadeh and Rudd (1986) (solid circles); Djuric, Cadez, and Kurepa (1988) (solid down triangles); Schutten et al. (1966) (solid up triangles); Khare and Meath (1987) (open down triangles); Straub et al. (1998) (open up triangles); and Olivero, Stagat, and Green (1972) (open diamonds). By comparison, the results obtained by Dingfelder et al. (1999) (open circles).

parameters, such as electron mobility and diffusion coefficient. Monte Carlo simulations have been performed by several groups that gave the electron elastic collision, ionization, and recombination rates.

3.3.2 Semiconductor Approximation

In considerations of charge transfer at the electrode interface, emphasis has traditionally been placed on field-controlled electron emissions. However, this approach is significantly compromised by the fast hydration of free electrons. In dielectric liquids, charged particles are normally present through the dissociative ionization of a neutral molecule:

$$M \leftrightarrow A^+ + B^- \tag{3.15}$$

The degree of the dissociative ionization depends highly on the short- and long-range binding forces. Usually, dissociations more readily occur in polar liquids (i.e., water) than nonpolar liquids because the high permittivity will weaken the Coulomb attraction between the anion and cation pairs. In liquid water, the proton H^+ combines with an H_2O molecule to form the H_3O^+ complex. This complex can transfer an excessive proton to the neighboring water molecule. This molecule turns into the H_3O^+ complex, and the transferring process can be passed on via the displacement of electrons, while the proton can be regarded as stationary. The charge carriers move under the strong electric field by the mechanism described, obtaining enough energy and eventually causing the ionization. This hopping process of the charge particles is similar to the hole mechanism in semiconductor physics and hence is called the semiconductor approximation (Lewis, 1994; Lewis, 1998).

For the gaseous state, the positive and negative ions will have energy levels of E_c and E_a, respectively, due to the loss/gain of the electrons. The cases in liquid are more complicated, mostly due to the solvation of the ions. The respective energy levels have to be modified by the collective polarization responses of the molecules surrounding the ions. The local reorganization of liquid molecules consists of two parts: the electronic component, which will grow on a time scale comparable with that needed to relocate the electron, and the atomic component, which can be considered frozen at the initial stage considering the Frank-Condon principle and will take a much longer time to form. The former causes the positive and negative ion energies (E_c and E_a) in the liquid to become $E_c - P$ and $E_a + P$, respectively, where P is the associated electronic solvation energy (as shown in Figure 3.7). The typical value for P is about 1 eV for nonpolar liquids, with relative permittivity of 2–3 and several electron volts for liquids with high dielectric permittivity (such as water, with $\varepsilon_r = 80$).

Considering the relative slowness of the atomic reorganization, the energy level will be further shifted by an energy amount of λ. Due to the different

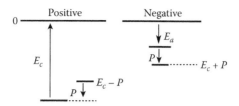

FIGURE 3.7
Electronic energy states for positive and negative molecular ions in gaseous phases.

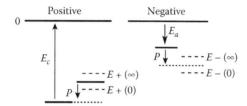

FIGURE 3.8
The localized $E(\infty)$ and delocalized $E(0)$ energy states of positive and negative ions in a liquid.

time required for P and λ shift, it is possible to define two electronic states for each ion. In the case of the negative ion state the energy is $E_-(0)$ at the moment when an electron is localized at a molecular site and $E_-(\infty)$ when the electron is in a fully polarized state after reorganization has occurred (Figure 3.8). Likewise, for a positive ion (hole) state, $E_+(0)$ is the energy of the hole or positive ion at the moment when an electron leaves a neutral molecular site, while $E_+(\infty)$ is the subsequent energy of the fully polarized positive ion state. The states between $E(0)$ and $E(\infty)$ correspond to various degrees of localization. Note that the edges of $E(0)$ and $E(\infty)$ are usually "blurred" by the thermal disturbance of the particles. The band-gap energy $[E_-(0) - E_+(0)]$ between the two sets of states is the necessary energy to create a quasifree electron-hole pair, as an analogy to the notions from semiconductor physics.

Under some circumstances (i.e., a high electric field), the electrons and holes that become localized in dielectric liquids can tunnel from one molecule to the next as shown in Figure 3.9. If tunneling occurs at an energy E in the band of states through an intermolecular potential barrier $V(a)$, assumed for simplicity to be one dimensional (Figure 3.9), then the probability of a transition depends on the following factor:

$$f = \exp\left[\frac{-4\pi}{h} \int_{x_1}^{x_2} [2m(V(x) - E)]^{\frac{1}{2}} dx\right] \tag{3.16}$$

where x_1, x_2 are the spatial limits of x where $V(x) = E$. The probability, hence the rate, of the transition depends strongly on both the height of the potential well and the distance between two molecules.

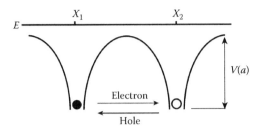

FIGURE 3.9
Transfer of electron by tunneling between molecules.

3.4 "Bubble Theory" for Electric Breakdown in Liquid

Although different electron avalanche mechanisms have been proposed to explain the electrical breakdown in liquid, the nature of the breakdown is still heavily disputed. In gas, the critical breakdown condition is described by the Paschen curve (Section 3.2.1), with which one can estimate the breakdown voltage for different gases at different pressures. For example, a value of 30 kV/cm is a well-accepted breakdown voltage of air at 1 atm. Due to the density difference between air and water, one could expect from the Paschen curve that the breakdown voltage in water would be approximately three orders of magnitude higher than that in air. However, many experimental data on the water breakdown showed that this breakdown voltage was higher but still of the same order of magnitude as for gases. In other words, the breakdown of liquids can be performed not at extremely high electric fields required by the Paschen curve but at those that only slightly exceed the breakdown electric fields in atmospheric pressure molecular gases. This interesting and practically important effect can be explained by taking into account the fast formation of gas channels in the body of water under the influence of an applied high voltage. When formed, the gas channels give the space necessary for the gas breakdown inside water, explaining why the voltage required for the breakdown in water is of the same magnitude as that in gases. The gas channels can be formed through the development and electric expansion of gas bubbles already existing in water or by additional formation of the vapor channel through fast local heating and evaporation.

3.4.1 Bubble Formation: Interface Processes

The influence of electrical charges on the surface is important to the physical chemistry of the surface. To begin, consider a solid metal surface bearing a uniform positive charge density in contact with a solution phase that contains both positive and negative ions. When the applied electric field is zero, charged layers will be established at the metal-liquid interface, mainly due to the contact potential between the metal and the liquid. The metal side of the interface will consist of an electron cloud extending beyond the positive cores of the metal surface. Under the effect of Coulombic force, the ions in the liquid will be attracted and adsorb chemically or physically on this layer to form the Stern and inner Helmholtz layers as shown in Figure 3.10. Farther outside are the outer Helmholtz layer and diffuse layer, where the net charge decays to zero in the bulk liquid and effectively screens the Stern and inner Helmholtz layer from the bulk of the solution. The structure of the electrical double layer is presented in Figure 3.10.

The formation of the so-called bubble or void is usually associated with surface tension γ at the interface. Under constant temperature and pressure conditions, the relationship between γ and the potential difference V across the interface can be described by the Lippmann capillary equation:

$$\frac{d\gamma}{dV} = -q \qquad (3.17)$$

where q is the surface charge density on the metal side of the interface. Analysis by Lewis (2003) showed that an applied electric field could alter the electrode interfacial double-layer field and thus change the cohesion of the liquid at the electrode boundary. Under electric stress as high as a breakdown threshold, the cohesion of the liquid can be much reduced or even lost altogether at the electrode surface, leading to the formation of the "void."

Another factor that may contribute to the breakdown process is the enhanced local field of the electrode double layer that induces the Lippmann process and will simultaneously control charge transfer across the electrode-liquid interface. The Auger process can provide both energetic electron injection at a cathode and efficient positive hole or ion injection at an anode (Lewis, 2003).

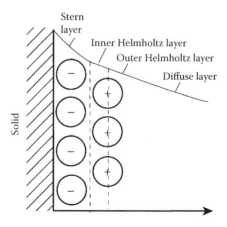

FIGURE 3.10
Schematic diagram of electrical double layer on the solid-liquid interface.

3.4.2 Bubble Formation: Joule Heating

Another possible mechanism for producing bubbles at the initial stage of electrical breakdown is through joule heating by ionic conduction inside liquid. The power of the joule heating per unit volume (W/m^3) is defined as

$$P = \sigma E^2 \tag{3.18}$$

where σ is the electric conductivity of the liquid, and E is the electric field. Usually, the energy from joule heating is well below the amount of energy required to heat the entire volume of liquid provided. However, considering that under normal breakdown conditions an electric field strength over 1 MV/cm is usually required, significant energy will be deposited through joule heating at the local region around the tip of an electrode, subsequently creating microbubbles near the electrode tip. If we take distilled water, for example, assume the electric conductivity is in the range of a few microsiemens per centimeter, then the time required for the generation of a gas bubble can be calculated to be a few hundred microseconds using the following properties of water: thermal capacity C_p = 4.186 kJ/kg/K, latent heat ΔH_{l-g} = 2,260 kJ/kg, $\mu E^2 \cdot \Delta t = \rho(C_p \Delta T + \Delta H_{l-g})$ with ΔT = 75 K. This rough estimation shows that internal heating can only be important if the duration of the applied voltage is sufficiently long. In recent experiments, the breakdown of water was observed with short subnanosecond voltage pulses. The time durations in these experiments were certainly several orders of magnitude shorter than that for any significant heating. Details about these nanosecond and subnanosecond breakdowns are discussed in further sections.

3.4.3 Bubble Formation: Preexisting Bubbles

The initiation of breakdown in water with the aid of preexisting bubbles of a size of 40–100 µm using pulse heated wire electrode was investigated by Korobeynikov et al. (2002). It was observed that the bubbles elongated in the direction of electric field, possibly due to the charge deposition on the gas-liquid interface. With a sufficiently high voltage, the discharge could be initiated in a bubble, with the prebreakdown time in the presence of a bubble much shorter than in its absence. Another interesting phenomenon is that the bubbles behave differently for negative and positive discharges. When a cathode bubble separates from an electrode surface, it does not participate in the streamer formation, indicating that direct cathodic electron injection in the bubble is of importance in this case.

Monte Carlo simulation was performed to show that a random distribution of preexisting microbubbles within the liquid would adequately explain the observed breakdown fields and time delays (Qian et al., 2006). Polarity differences in hold-off voltages and breakdown initiation times are also in agreement with the microbubble models.

3.5 Streamer Propagation

Despite different mechanisms proposed, all the initiation theories lead to the formation of a low-density region so that a self-sustained electron avalanche could be possible. Thus, the next question is what the driving force is to sustain and expand the cavity to form complex geometrical structures. Similar to the initiation process, the propagation is complicated because it involves interactions between plasma, gas, and liquid phases of the media. Recent experiments demonstrated the existence of different modes of propagation; both a primary streamer mode with a slow velocity and a secondary streamer mode with a high velocity were observed. Several models have been proposed to correlate electric field to streamer velocity. Different effects, including liquid viscosity, trapping of positive and negative carriers in the conducting channel, and local electric charge at the streamer head were taken into account. Again, there is not yet a commonly accepted model.

In the following section, we describe a theoretical framework for understanding the propagation of streamers of electric discharge in water subjected to high voltage. The breakdown process is usually characterized by two typical features of breakdown: rapid propagation of discharge streamers and high tendency of branching and formation of random dendritic structures. Therefore, the present theoretical review consists of two components: a quantitative model for possible mechanisms to produce the driving force needed to sustain and promote the propagation and stability analysis of a single cylindrical filament with surface charges in an external electric field.

3.5.1 Electrostatic Model

A schematic diagram of the present electrostatic model is shown in Figure 3.11a. A thin needle electrode with a rounded tip was aligned perpendicular to a ground-plate electrode. High voltage Φ_0 was applied on the needle electrode. According to Kupershtokh and Medvedev (2006), liquids

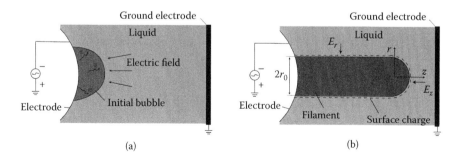

(a)   (b)

FIGURE 3.11
(a) Initiation of bubble formation; (b) schematic diagram of a cylindrical filament in water.

could become phase unstable under a high electric field strength, so gas channels could form along electric field lines. The time required for breakdown ignition in the channels can be estimated as $\tau_b = (k_i n_0)^{-1}$, where k_I is the direct ionization rate coefficient, and n_0 is the molecule density. Under atmospheric pressure, n_0 is on the order of 10^{19} cm^{-3}, while k_I is on the order of 10^{-10} to 10^{-9} cm^3/s in the reduced electric field E/n_0 of 10^3 V·cm^2. Hence, τ_b is on the order of 0.1 to 1 ns. For negative discharges, due to the higher momentum transfer collision frequency and thus a low mobility in the liquid phase, electrons tend to deposit on the gas-liquid interface and charge it negatively. For positive discharges, the high mobility of electrons would leave the interface charged positively. Under both circumstances, it is possible that the charged interface would be pushed to displace the liquid under external electric field by electrostatic force.

A simplified calculation can be made to examine whether the electrostatic force would be sufficient to overcome the resistance of water at the interface. The pressure due to the surface tension γ on a water interface of a spherical bubble with a radius of curvature r can be approximated by the Young-Laplace equation $p = 2\gamma/r$. With $r \sim 1$ μm and $\gamma = 72.8 \times 10^{-4}$ N/m, the surface tension pressure is about 15 kPa. The ultimate strength of water of approximately 30 MPa must be exceeded for rupturing the liquid. Considering forces due to charged particles only and ignoring those due to field gradients and material property gradients, the electric force at the interface becomes simply the electrostatic force L, which is the product of charge density per unit area σ and the electric field E, that is, $L = e\sigma E$, where e is the charge per electron. For $E = 10^8$ V/cm, σ has a value of 10^{12} charges/cm^2. For electrons with an average energy of 1 eV, the electron thermal velocity can be estimated as 6×10^7 cm/s. So, a modest electron density of 10^{13} cm^{-3} will provide the flux necessary to charge the surface to the breaking point within 1 ns. Although these estimations for water rupturing also neglect both loss mechanisms and the energy requirements to overcome the hydrodynamic resistance, the electrostatic mechanism still seems a likely candidate for streamer propagation, and such forces may dominate at a nanosecond time scale.

The growth of a plasma filament is determined by conservation equations of mass, momentum, and energy. To quantify the breakdown process described, the equations for the formation and propagation of the plasma-filled filaments are defined as (Gidalevich and Boxman, 2006)

$$\frac{\partial \rho}{\partial t} + \nabla \cdot (\rho u) = \frac{2\lambda(T)T}{\Delta_v H r_0^2} \tag{3.19}$$

$$\frac{\partial u}{\partial t} + u \cdot \nabla u + \frac{1}{\rho} \nabla P = 0 \tag{3.20}$$

$$\frac{\partial}{\partial t}\left(\rho(Z + u^2)\right) + \nabla \cdot \left(\rho u \left(Z + \frac{P}{\rho} + \frac{u^2}{2}\right)\right) = \kappa(T)E^2 \tag{3.21}$$

where t is time; ρ and P are the radial density and pressure inside streamer, respectively; u is the velocity of streamer; λ is the temperature; l is the thermal conductivity; $\Delta_v H$ is the evaporation heat of water; r_0 is the radius of streamer; Z is the internal energy of ionized gas; E is the electric field strength; and σ is the electric conductivity. It is usually difficult to directly solve Equations 3.19 to 3.21 because of the high nonlinearity of the equations.

For simplification, the streamer is assumed to be a cylinder with a hemispherical tip as shown in Figure 3.11b. The reference frame is fixed on the tip. The radius of the filament is r_0. Although it appears from photographic evidence that the filament is usually of a conical shape, the cylindrical approach is still a good approximation when the length of the filament is much greater than the radius. The electric conductivity σ inside the filament could be described as

$$\sigma = \frac{n_e e^2}{m v_{en}} \tag{3.22}$$

where m is the mass of electron, and v_{en} is the frequency of electron-neutral collisions. Note that v_{en} is proportional to the gas number density, and the value of v_{en}/p is usually on the order of 10^9 s^{-1} Torr^{-1}. Sunka et al. (1999) measured the broadening of the Hα line profile, which is commonly used to characterize the density of plasma, reporting the electron density inside streamers during the initial phase of water breakdown, to be on the order of 10^{18} cm^{-3}. With saturated water vapor pressure of 20 Torr at room temperature, the electric conductivity inside the filament can be estimated to be on the order of 10^7 S/m, a value comparable to those for metals. So, the filament could be regarded as equipotential with the electrode and thus could be treated as an extension of the electrode throughout the expansion. The external fluid provides drag force and constant external pressure for the development of the filament. Gravity is neglected here because the body force induced by gravity is much smaller than electric forces.

The electric field outside a slender jet can be described as if it were due to an effective linear charge density (i.e., incorporating effects of both free charge and polarization charge) of charge density ρ_e on the surface. Since the charge density in liquid can be ignored compared with that on the filament surface, one can have the following equation for the space outside the filament by applying the Laplace equation in the radial direction:

$$\frac{1}{r} \frac{\partial}{\partial r} \left(r \frac{\partial \Phi}{\partial r} \right) = 0 \tag{3.23}$$

with boundary condition: $\Phi|_{r=r_0} = \Phi_0$ and $\Phi|_{r=R} = 0$. R is the distance between anode and cathode. Since the filament could be regarded as an extension of the electrode, R decreases as the streamer propagates through the gap.

Solving Equation 3.23 with an assumption of negative discharge, the radial electric field E_r and local surface charge density σ_r can be written as

$$E_r = \frac{\partial \Phi}{\partial r} = -\frac{\Phi_0}{r_0 \ln(R/r_0)} \tag{3.24}$$

$$\rho_{er} = \varepsilon E_{r_0} = -\varepsilon_r \varepsilon_0 \frac{\Phi_0}{r_0 \ln(R/r_0)} \tag{3.25}$$

Since there is no analytical solution for the electric field at the hemispherical tip of the filament, a frequently used approximation is $E_z \approx \Phi_0/r_0$. Here, the equation for the electric field at the tip of a needle in a needle-to-plane geometry developed by Lama and Gallo (1977) was used:

$$E_z = -\frac{2\Phi_0}{r_0 \ln(4R/r_0)} \tag{3.26}$$

Similarly, the local charge density at the tip can be given as

$$\rho_{ez} = \varepsilon E_z = -\varepsilon_r \varepsilon_0 \frac{2\Phi_0}{r_0 \ln(4R/r_0)} \tag{3.27}$$

From Equations 3.24 to 3.27, one can conclude that the radial direction electrostatic pressure $E\sigma$ exerted on the side wall of the streamer is weaker than the axial direction electrostatic pressure on the tip. Note that both electrostatic pressures are roughly inversely proportional to r_0^2, meaning that at the initial stage of the filament growth when r_0 is small, the electrostatic forces on both directions are strong, and the filament will grow both axially and radially. A direct consequence of both the axial and radial expansions of the streamer channel is the launching of compression waves into adjacent liquids (An, Baumung, and Bluhm, 2007). At some critical point, the electrostatic force will reach a balance with hydrodynamic resistance acting on the surface in the radial direction first, while the filament continues to grow in the axial direction.

Experimentally recorded propagation speeds of the filaments varied depending on the measurement techniques, ranging from a few kilometers per second to 100 k/s. In spite of the discrepancy observed by different groups, the propagation was clearly in the supersonic regime, and the formation of shock waves had to be taken into consideration (see Figure 3.12). The drag force on the tip of the streamer, which is a stagnation point, equals the force produced by the total hydrodynamic pressure:

$$P_{hd} = P_1 \left(\frac{2\alpha}{\alpha+1} M_1^2 - \frac{\alpha-1}{\alpha+1} \right) + \frac{1}{2}\rho(CM_2)^2 \tag{3.28}$$

where P_1 is ambient pressure, $P_1 \cdot [(2\alpha/\alpha + 1)M_1^2 - (\alpha - 1/\alpha + 1)]$ is the pressure behind the shock front, α is the specific heat ratio of water, M_1 is the Mach

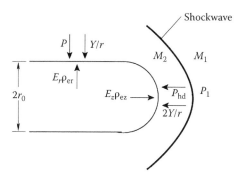

FIGURE 3.12
Force balance for the present electrostatic model.

number of streamer, M_2 is the Mach number after the shock front, and C is the speed of sound in liquid. The relationship between M_1 and M_2 can be written as (White, 2006)

$$M_2^2 = \frac{(\alpha - 1)M_1^2 + 2}{2\alpha M_1^2 + 1 - \alpha} \tag{3.29}$$

Equating the hydrodynamic pressure to the sum of the electrostatic pressure and the pressure produced by surface tension at the tip, one can have the following equation for streamer propagation:

$$4\varepsilon_r \varepsilon_0 \frac{\Phi_0^2}{r_0^2 \ln^2(4R/r_0)} = P_1 \left(\frac{2\alpha}{\alpha + 1} M_1^2 - \frac{\alpha - 1}{\alpha + 1} \right) + \frac{1}{2}\rho(CM_2)^2 + \frac{2\gamma}{r_0} \tag{3.30}$$

The balance between the electrostatic force and the force produced by the total hydrodynamic pressure in the radial direction can be given as

$$\varepsilon_r \varepsilon_0 \frac{\Phi_0^2}{r_0^2 \ln^2(R/r_0)} = P_1 + \frac{1}{2}\rho(CM_2)^2 + \frac{\gamma}{r_0} \tag{3.31}$$

Note that there are three unknowns, M_1, M_2, and r_0 in the preceding equations. So, it is possible to solve Equations 3.29 to 3.31 simultaneously when the applied voltage Φ_0 and the interelectrode distance R are specified.

To demonstrate the validity of the present model, the filament radius predicted by the model is shown in Figure 3.13. For a typical interelectrode distance of 1 cm, the filament radius increased from 3 to 50 µm as the applied voltage rose from 5 to 30 kV. The value was comparable to typical experimental values. For example, An, Baumung, and Bluhm (2007) reported that the light emission from the discharge was restricted to a channel of 100–µm diameter, indicating the interaction of charged particles in the region.

Figure 3.14 shows the filament propagation speed as a function of Φ_0 and R. The calculated propagation speed from the present model was around 15

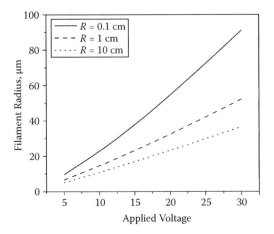

FIGURE 3.13
Variations of filament radius as a function of applied voltage and interelectrode distance.

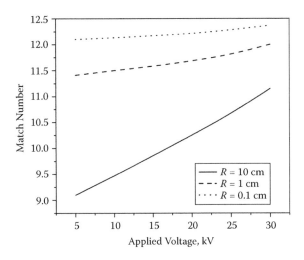

FIGURE 3.14
Variations of the Mach number of streamer as a function of applied voltage and interelectrode distance.

km/s, which was higher than the primary streamer speed but lower than the secondary streamer speed reported by An, Baumung, and Bluhm (2007). The Mach number increased moderately with the applied voltage, a phenomenon that was understandable from the point of view of energy conservation. The streamer propagation velocity was relatively independent of the interelectrode distance. For an applied voltage of 30 kV, the Mach number increased from 11.2 to 12.3 when interelectrode distance decreased from 10 to 0.1 cm. This was consistent with the known property of negative streamers as the previous experiment showed that, for a given voltage, the propagation

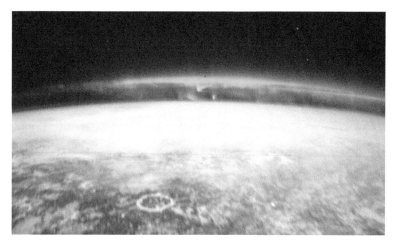

FIGURE 1.1
Aurora borealis as seen from International Space Station. (From NASA.)

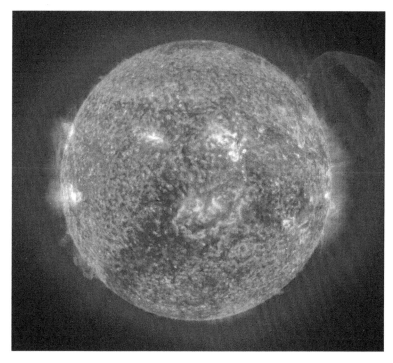

FIGURE 1.2
Solar plasma. Emission in spectral lines shows the upper chromosphere at a temperature of about 60,000 K. (From NASA.)

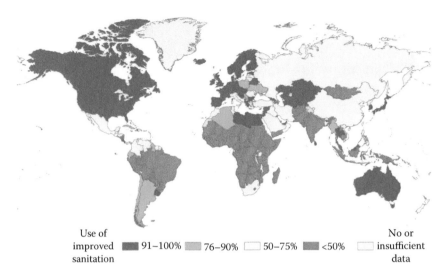

Use of improved sanitation ■ 91–100% ■ 76–90% ☐ 50–75% ■ <50% No or ☐ insufficient data

FIGURE 1.4
Use of sanitary water in rural areas, 2008. (From World Health Organization. (2010) Progress on sanitation and drinking-water.)

FIGURE 1.5
Sandia National Lab's Z machine, bathed in transformer oil and deionized water for greater electric insulation. (Courtesy of Sandia National Lab, 2004.)

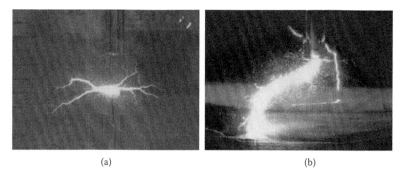

FIGURE 2.2
Images of plasma discharge in water: (a) pulsed corona; (b) pulsed arc.

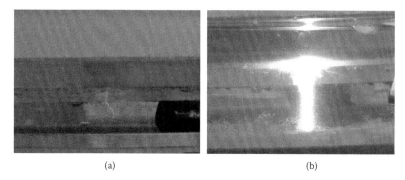

FIGURE 2.5
Images of plasma discharges through a pinhole: (a) pulsed corona; (b) pulsed arc produced at Drexel Plasma Institute.

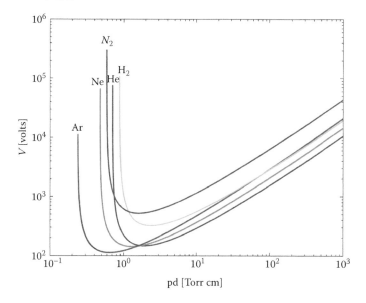

FIGURE 3.2
Paschen curves for different gases. (From Fridman, A. (2008) *Plasma chemistry*, Cambridge University Press, Cambridge.)

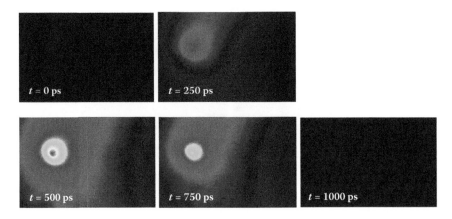

FIGURE 3.23
Discharge emission dynamics. 1,000 accumulations/image. Camera gate was 500 ps, $\Delta\lambda$ = 250–750 nm. Image size was 7×4 mm.

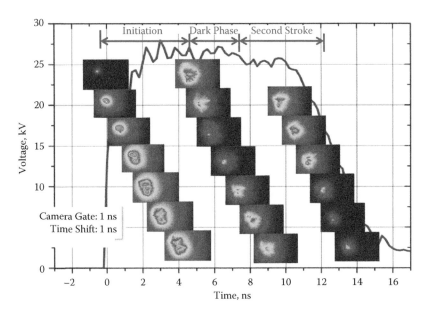

FIGURE 3.24
Dynamics of discharge emission and high-voltage potential on electrode. Distilled water, U = 27 kV. Camera gate is 1 ns. Spectral response 250–750 nm.

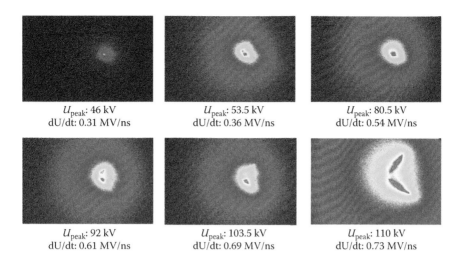

FIGURE 3.25
Discharge development for different voltage. Pulse width was 400 ps. ICCD camera gate was 100 ns.

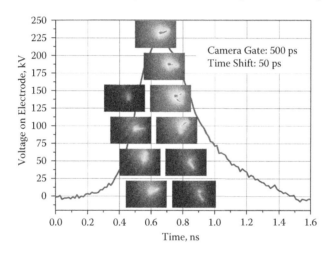

FIGURE 3.26
Dynamics of discharge emission. Distilled water. U = 220 kV. dU/dt = 1.46 MV/ns. Camera gate was 500 ps. Time shift between frames was 50 ps.

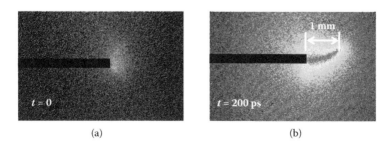

FIGURE 3.27
Velocity of streamer propagation in water. 150 ps rise time, 1.5 MV/ns: Growth rate $v_z = L/\Delta t$ ≈ 1 mm/200 ps = 5,000 km/s ~ 15% c. Expansion rate $v_r = r/\Delta t$ ≈ 0.05 mm/200 ps = 250 km/s.

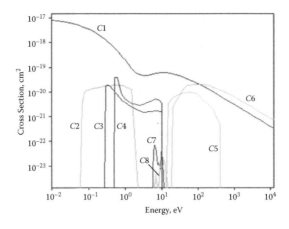

FIGURE 3.28
Excitation cross section (m²) for H_2O vapor (rarefied gas). C_1, elastic collisions; C_2, rotational excitation (0.02 eV); C_3, vibrational excitation (0.20 eV); C_4, vibrational excitation (0.45 eV); C_5, electronic excitation (7.10 eV); C_6, ionization (12.61 eV); C_7, C_8, dissociative attachment.

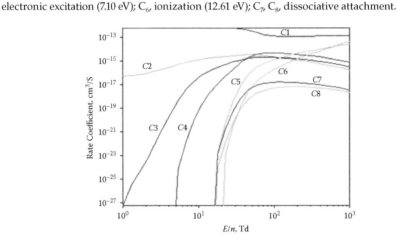

FIGURE 3.29
Rate coefficients (m³/s) for different processes in water vapor in dependence on the E/n value. C_1, transport collisions; C_2, rotational excitation; C_3, C_4, vibrational excitation; C_5, electronic excitation; C_6, ionization; C_7, C_8, dissociative attachment.

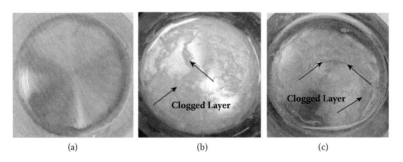

(a)   (b)   (c)

FIGURE 5.13
The clotted formation of white layer in blood plasma sample with DBD treatment: (left) control; (middle) blood plasma treated with DBD for 4 min; (right) blood plasma treated with DBD for 8 min.

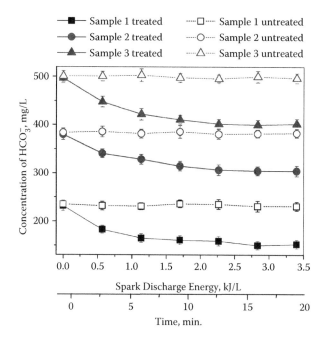

FIGURE 6.9
Variations of HCO_3^- over time for cases with and without plasma treatment. (From Yang, Y., Kim, H., Starikovskiy, A., Fridman, A., and Cho, Y.I. (2010) Application of pulsed spark discharge for calcium carbonate precipitation in hard water. *Water Res.* 44, 3659–3668.)

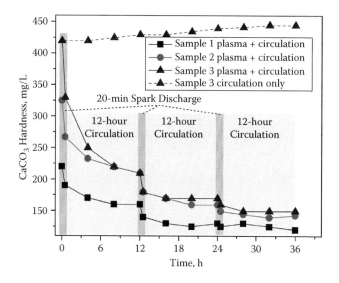

FIGURE 6.14
Variations of (a) $CaCO_3$ hardness and (b) pH value over time with plasma treatment and spray circulation.

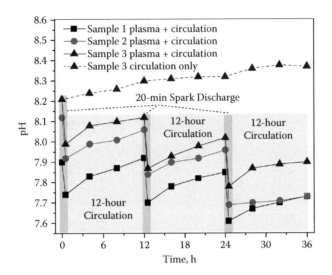

FIGURE 6.14 (CONTINUED)
Variations of (a) CaCO₃ hardness and (b) pH value over time with plasma treatment and spray circulation.

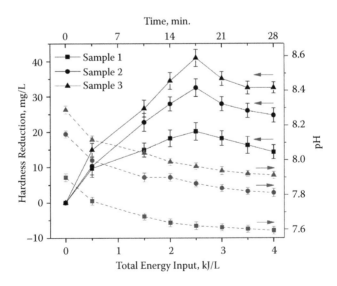

FIGURE 6.22
Variations of calcium carbonate hardness and pH over time for different energy inputs by the transient hot-wire method. (From Yang, Y., Kim, H., Starikovskiy, A., Fridman, A., and Cho, Y.I. (2011b) Precipitation of calcium ions from hard water using pulsed spark discharges and its mechanism. *Plasma Chem. Plasma Proc.* 31, 51–66.)

velocity was relatively constant as the streamer crossed the gap, while it increased as the streamer approached the plane electrode. This phenomenon could be understood by Equation 3.24: The interelectrode distance R was decreased with the propagation of the streamers; as a result, the electric field at the tip of the streamer was increased, leading to a higher propagation speed. However, the amount of the increase in the electric field would not be significant because of the natural logarithm in Equation 3.31.

3.5.2 Thermal Mechanism

In the electrostatic model described, it was assumed that the translational temperature inside the streamer was low, and the electrostatic force was the only driving force for the growth of the filament. The assumption was valid only at the initial stage of the filament development as the temperature kept rising as the molecules gained more energy through electron-neutral collisions. The heating time τ was approximately $\tau = \tau_{en} + \tau_{vt}$, where τ_{en} was the time for electron-neutral excitation, and τ_{vt} was the time for vibrational-translational (v-t) relaxation. For electron-neutral excitation, $\tau_{en} = 1/\nu_{en} = 1/(n_e k_{en})$, where ν_{en} is the electron-neutral nonelastic collision frequency, n_e is the electron density, and k_{en} is the rate constant for electron-neutral collisions. Then, k_{en} can be expressed as $k_{en} = s_{en} v_{te}$, where s_{en} is the cross section for vibrational excitation of H_2O molecules by electron impact, and v_{te} is the electron thermal velocity.

For electrons with an average energy of 1 eV, the cross section for vibrational excitation is about $\sigma = 10^{-17}$ cm^2. Thus, k_{en} is about 10^{-8} cm^3/s as is typical ($v_{te} = 6 \times 10^7$ cm/s). Spectroscopic measurements indicated that the stark broadening of Hα line corresponded to an electron density of about 10^{18} cm^{-3} at a quasiequilibrium state. Thus, the typical electron-neutral excitation time can be estimated as on the order of a few nanoseconds. For the v-t relaxation, $\tau_{vt} = 1/(n_v k_{vt})$, where n_v is the density of vibrational excited molecules, and k_{vt} is the v-t relaxation rate coefficient. For water molecules at room temperature, k_{vt} is about 3×10^{-12} cm^3/s (Fridman and Kennedy, 2006). Assuming that n_v is in the same order with electron density, τ_{vt} could be estimated as on the order of several hundred nanoseconds, suggesting that heating can take place inside the filaments under a submicrosecond time scale due to the energy transfer from the electrons to the translational energy of the water molecules; furthermore, the propagation of the streamers could be caused by the continuous evaporation of water molecules at the tip. Here, the energy dissipation was not considered, and the actual heating time might be longer, but still the local heating mechanism under the submicrosecond time regime seems possible.

To quantify the process described, it was assumed that a small cylindrical portion of water evaporated at the tip of the streamer during time Δt so that the length of the streamer grew from L to $L + \delta L$, as shown in Figure 3.15. The diameter of the evaporated water cylinder was assumed to be $2r_e$. There was

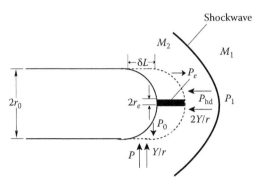

FIGURE 3.15
Force balance for the present thermal model.

no definitive value for pressure P_e inside the small vaporized portion given the extremely high temperature. However, P_e could be estimated to be on the order of 1,000 atm because of the density difference between liquid water and vapor. Such a high pressure could provide the driving force needed for the growth of the filament. As in the previous section, one could get the force balance along the axial direction at the tip of the filament assuming a steady-state condition as

$$P_1\left(\frac{2k}{k+1}M_1^2 - \frac{k-1}{k+1}\right) + \frac{1}{2}\rho(CM_2)^2 + \frac{2\gamma}{r_e} = P_e \tag{3.32}$$

The energy required for the evaporation of water can be calculated as

$$E_e = \rho V_e(c_p\Delta T + \Delta_v H) \tag{3.33}$$

where ρ is the density of water, V_e is the volume of evaporated water, and C_p is the specific heat of water. V_e can be written as

$$V_e = \pi r_e^2 \cdot \delta L \tag{3.34}$$

After evaporation, the overheated and overpressured water vapor will expand radially, while satisfying the force balance along the axial direction, until it reaches equilibrium with the outside hydrodynamic pressure. The process could be regarded as adiabatic under a submicrosecond time scale, and thus one can have the following equations:

$$P_e V_e^{\alpha_s} = P_0 V_0^{\alpha_s} \tag{3.35}$$

$$\frac{P_0}{P_e} = \left[\frac{\varepsilon_r\varepsilon_0\Phi_0^2}{\rho r_0(c_p\Delta T + \Delta H_{l-g})\ln(R/r_0)}\right]^{\alpha_s} \tag{3.36}$$

where P_0 and V_0 are the pressure and volume, respectively, of the water vapor after the expansion; r_0 is the radius of the filament after (Figure 3.16)

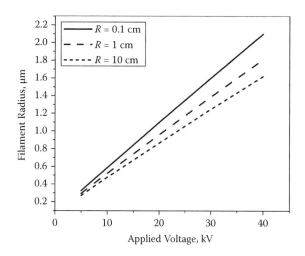

FIGURE 3.16
Variations of filament radius as a function of applied voltage and interelectrode distance.

expansion; and α_s is the specific heat ratio of water vapor. The force produced by P_0 should be in balance with the forces created by both surface tension and total environmental hydrodynamic pressure as follows:

$$P_0 = P_1 + \frac{1}{2}\rho(CM_2)^2 + \frac{\lambda}{r_0} \tag{3.37}$$

Another set of equations can be obtained through the consideration of energy conservation. The energy required to vaporize water was the electric energy provided by the power supply. If the entire filament was viewed as a capacitor with capacitance C, the required energy could be calculated as

$$E = \frac{C\Phi_0^2}{2} \tag{3.38}$$

The capacitance of the cylindrical filament is

$$C = 2\pi\varepsilon\varepsilon_0 r_0 L / \ln(R/r_0) \tag{3.39}$$

So, the energy change required to extend the length by δL becomes

$$\delta E = \pi\varepsilon\varepsilon_0 r_0 \delta L \Phi_0^2 / \ln(R/r_0) \tag{3.40}$$

By equating δE to E_e, one has

$$r_e = \sqrt{\frac{\varepsilon\varepsilon_0 r_0 \Phi_0^2}{\rho(c_p \Delta T + \Delta H_{l-g})\ln(R/r_0)}} \tag{3.41}$$

Assuming $\alpha \approx 1$ due to the low compressibility of water, and rearranging Equations 3.29 3.32, 3.36, 3.37, and 3.41 to eliminate M_2, r_e, and P_g, one can get a set of equations about M_1 and r_0 as follows:

$$P_1 M_1^2 + \frac{\rho C^2}{2M_1^2} - P_e + 2\gamma \sqrt{\frac{\rho(c_p \Delta T + \Delta H_{l\text{-}g})\ln(R/r_0)}{\varepsilon_r \varepsilon_0 r_0 \Phi_0^2}} = 0 \qquad (3.42)$$

$$P_e \left[\frac{\varepsilon_r \varepsilon_0 \Phi_0^2}{\rho r_0 (c_p \Delta T + \Delta H_{l\text{-}g})\ln(R/r_0)} \right]^{k_s} - P_1 - \frac{\rho C^2}{2M_1^2} - \gamma \sqrt{\frac{\rho(c_p \Delta T + \Delta H_{l\text{-}g})\ln(R/r_0)}{\varepsilon_r \varepsilon_0 r_0 \Phi_0^2}} = 0$$

$$(3.43)$$

For water vapor, α_s can be assumed to be 1.3 (Parry et al., 2008). For high-temperature underwater discharges, the translational plasma temperature was measured to be between 4,000 and 6,500 K (Lange and Huczko, 2004). An average value of 5,000 K was used for ΔT in the present study. Figure 3.17 shows the Mach number of filament propagation M_1 as a function of Φ_0 and R. The propagation velocity was about 50 km/s, which was higher than the secondary streamer velocity of 25 km/s reported by An, Baumung and Bluhm (2007), but lower than the value of 200 km/s reported by Woodworth et al. (2004). The discrepancy in the two measurements probably came from the different techniques used for the velocity measurements. The value of M_1 remained constant for various values of Φ_0 and R, indicating that the propagation velocity of the streamers was independent of either the applied voltage or interelectrode distance. A similar phenomenon was observed previously; the propagation velocity of secondary streamers was constant over a wide voltage range. Figure 3.16 shows the filament radius as a function of Φ_0

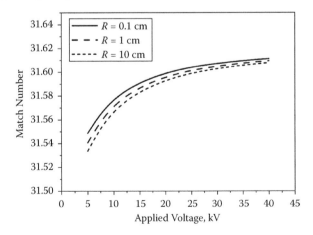

FIGURE 3.17
Variations of the Mach number of streamer as a function of applied voltage and interelectrode distance.

and R. The radius increased slightly as the streamers approached the other electrode, while it decreased almost linearly as the applied voltage dropped. The absolute value of r_0 was about one order smaller than that obtained from the electrostatic model. This can be understood if one considers the energy requirements for the two mechanisms. For the evaporation of water, the energy needed to break the hydrogen bonds between water molecules should be much greater than that required to displace the same volume of water.

The different models based on the electrostatic force and evaporation gave different results of the streamer propagation speed and filament radius. The electrostatic model showed streamers with a larger radius and a lower Mach number, while the thermal model demonstrated that the streamers could move much faster but were thinner than those determined from the electrostatic model. The different findings from the two models led us to postulate that different mechanisms might be associated with the different modes of the streamer propagation. At the initial primary streamer mode before any significant heat was generated, the electrostatic force might have played a major role. The appearance of the secondary streamer required more time, during which the electron energy could be transferred to translational energy of water molecules and subsequently evaporation become the dominant force to drive the filament to move forward. The transition time between the primary and secondary streamers was on the order of 100 ns, a value in accordance with the heating time as estimated.

3.6 Stability Analysis of the Streamers

The breakdown process is usually characterized by two features: an initial development of thin discharge streamer channels and a subsequent branching of these channels into complicated "bushlike" patterns (Takeda et al., 2008). Apparently, the branching process is associated with the instability of the filament. In this section, the linear stability analysis of axisymmetric perturbation of a filament surface with a certain electric charge density is presented. As long as the wavelength of the perturbation is much smaller than the length of the filament, the stability characteristics can be approximated by considering perturbations to a charged cylinder of constant radius as shown in Figure 3.18. The peak-to-peak amplitude and wave number of the disturbance are h and k, respectively. H is the depth of wave influence, and u is the velocity of liquid relative to the disturbance. Then, the surface of the perturbation can be represented by the following equation:

$$r = r_0 + \frac{h}{2}\exp(ikz + \omega t) \tag{3.44}$$

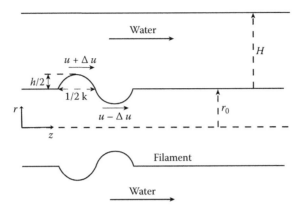

FIGURE 3.18
Schematic diagram of disturbance at the surface of filament.

where ω is the oscillation frequency of the instability.

To analyze the linear stability, the disturbance of the local electrostatic force, surface tension, and hydrodynamic pressure were considered following a geometrical perturbation. Generally, the surface tension tends to minimize the surface area and subsequently stabilize the disturbance, while the local enhancement of the electrostatic force tends to push the disturbance to grow. In the reference frame that moves together with the tip of filament, the effects of these three forces were considered separately for the pressure balance between the crest and trough along the stream line (see Figure 3.18).

3.6.1 Electrostatic Pressure

According to Equations 3.24 to 3.27, the electrostatic pressure is proportional to the square of the local curvature of the interface, which is different at the crest and trough of the perturbation. Thus, the electrostatic pressures at the crest and trough $P_{E,c}$ and $P_{E,t}$ become, respectively,

$$P_{E,c} = \varepsilon_r \varepsilon_0 \Phi_0^2 \frac{\chi_c^2}{4} \tag{3.45}$$

and

$$P_{E,t} = \varepsilon_r \varepsilon_0 \Phi_0^2 \frac{\chi_t^2}{4} \tag{3.46}$$

where ε_r is the relative permittivity of water, and χ_c and χ_t are the mean curvatures at the crest and trough, respectively. The expression for the mean curvature can be written as

$$\chi = \frac{1}{r\sqrt{1+(\partial_z r)^2}} - \frac{\partial_z(\partial_z r)}{(1+(\partial_z r)^2)^{3/2}} = \frac{1}{r} - \partial_z(\partial_z r) \tag{3.47}$$

Substituting Equation 3.44 into Equation 3.47, one can obtain expressions for M_c and M_t:

$$\chi_c = \frac{1}{r_0 + h/2} + \frac{h}{2}k^2 \tag{3.48}$$

$$\chi_c = \frac{1}{r_0 + h/2} - \frac{h}{2}k^2 \tag{3.49}$$

Subsequently, $P_{E,c}$ and $P_{E,t}$ can be written as

$$P_{E,c} = \varepsilon_r\varepsilon_0\Phi_0^2\frac{\chi_c^2}{4} = \varepsilon_r\varepsilon_0\Phi_0^2\left(\frac{1}{(2r_0+h)^2} + \frac{hk^2}{2(2r_0+h)}\right) \tag{3.50}$$

$$P_{E,t} = \varepsilon_r\varepsilon_0\Phi_0^2\frac{\chi_c^2}{4} = \varepsilon_r\varepsilon_0\Phi_0^2\left(\frac{1}{(2r_0-h)^2} - \frac{hk^2}{2(2r_0-h)}\right) \tag{3.51}$$

Thus, the electrostatic pressure difference between the crest and trough becomes

$$\Delta P_E = P_{E,c} - P_{E,t} = -\frac{\varepsilon_r\varepsilon_0\Phi_0^2 h}{2r_0^3} + \frac{\varepsilon_r\varepsilon_0\Phi_0^2 hk^2}{2r_0} \tag{3.52}$$

3.6.2 Surface Tension

The pressures due to the surface tension across the interface at the crest and trough can be written as

$$P_{T,c} = \gamma\chi_c = \gamma\left(\frac{1}{r_0+h/2} + \frac{h}{2}k^2\right) \tag{3.53}$$

and

$$P_{T,t} = \gamma\chi_t = \gamma\left(\frac{1}{r_0-h/2} - \frac{h}{2}k^2\right) \tag{3.54}$$

Thus, the pressure difference due to surface tension between the crest and trough becomes

$$\Delta P_T = P_{T,c} - P_{T,t} = -\frac{\gamma h}{r_0^2 - h^2/4} + \gamma hk^2 \tag{3.55}$$

Since $r_0 \gg h$, Equation 3.54 can be simplified as

$$\Delta P_T = -\frac{\gamma h}{r_0^2} + \gamma hk^2 \tag{3.56}$$

3.6.3 Hydrodynamic Pressure

When there is a disturbance on the interface of the filament, the flow speed of liquid will be perturbed in the depth of wave influence, inducing a hydrodynamic pressure difference ΔP_H between the crest and trough:

$$\Delta P_{HD} = \frac{1}{2}\rho\left(u + \frac{\Delta u}{2}\right)^2 - \frac{1}{2}\rho\left(u - \frac{\Delta u}{2}\right)^2 = \rho u \Delta u \qquad (3.57)$$

where $\Delta u/2$ is the perturbation in the flow speed caused by the shape of the wave. The dynamic pressure is related to the flow speed through Bernoulli's equation. The pressure difference from the electrostatic force and dynamic effect of the flow has the opposite sign due to the surface tension. For a balance between two kinds of oppositely directed pressure differences, one has

$$\rho u \Delta u + \frac{\gamma h}{r_0^2} - \gamma h k^2 - \frac{\varepsilon_r \varepsilon_0 \Phi_0^2 h}{2r_0^3} + \frac{\varepsilon_r \varepsilon_0 \Phi_0^2 h k^2}{2r_0} = 0 \qquad (3.58)$$

To solve Equation 3.58, the perturbed flow speed Du must be expressed in terms of experimentally measurable quantities. The derivation that follows was inspired by Kenyon (1983; 1998).

Assuming that the perturbed flow speed is constant over the depth of wave influence, the mass conservation equation through vertical cross sections between the crest and trough becomes

$$\left(u + \frac{\Delta u}{2}\right)\left(H - \frac{h}{2}\right) = \left(u - \frac{\Delta u}{2}\right)\left(H + \frac{h}{2}\right) \qquad (3.59)$$

where H is the depth of wave influence. Equation 3.59 can be reduced to

$$u \cdot h = \Delta u \cdot H \qquad (3.60)$$

The theoretical expression for H was given by Kenyon as

$$H = \frac{1}{2\pi k} \qquad (3.61)$$

Using Equations 3.60 and 3.61 to eliminate H and Δu, Equation 3.58 becomes

$$\rho \omega^2 = k\left(\gamma k^2 + \frac{\varepsilon_r \varepsilon_0 \Phi_0^2}{2r_0^3} - \frac{\gamma}{r_0^2} - \frac{\varepsilon_r \varepsilon_0 \Phi_0^2 k^2}{2r_0}\right) \qquad (3.62)$$

Since this is a quadratic equation, there will be two different branches of the dispersion relation, and an instability occurs if $Re(\omega) > 0$. The first thing to note in Equation 3.62 is that when the applied voltage Φ_0 is equal to zero and the surface is flat, in other words, when the radius of the filament r_0 goes to

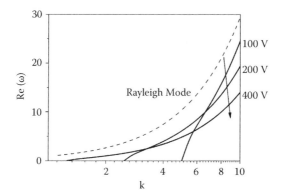

FIGURE 3.19
Instability growth rate ω at low applied voltages. k and ω are nondimensionalized using streamer radius r_0 and time scale $t = (\rho r_0^3/\gamma)^{1/2}$.

infinity, Equation 3.62 reduces to $\rho u^2 = \gamma k$, which is the equation for the classic two-dimensional Rayleigh instability (Eggers, 1997).

Figure 3.19 shows the instability growth rate ω at a low applied voltage, where the process is in Rayleigh mode. The dashed line represents the classic Rayleigh instability for $\Phi_0 = 0$ and $r_0 \to \infty$. For $\Phi_0 \neq 0$ and r_0 is finite, instability only happens at high wave numbers. When the voltage increases under this mode, the growth rate is decreased until fully suppressed at a certain critical value. The physical explanation for this can be as follows: The Rayleigh instability occurs due to surface tension, which always acts to break a cylindrical jet into a stream of droplets; on the other hand, the electrostatic force, which is proportional to the square of the applied voltage, always acts on the opposite direction of the surface tension. When the applied voltage increases, the Rayleigh instability would be suppressed when the two forces are balanced.

As the voltage continues to increase, the instability enters the electrostatic mode, where the electrostatic force exceeds the force created by the surface tension and becomes the dominant force. Figure 3.20 shows the instability growth rate ω at a high voltage. Both the growth rate and the wave number range increase as the voltage rises. The physics of this mode is a consequence of the interaction of the electric field with the surface charge on the interface; surface tension is a parameter of less importance for this mode. The mechanism for the instability is that a perturbation in the radius of the filament induces a perturbation in the surface charge density and therefore a perturbation in the electrostatic pressure. At a high voltage, the perturbation is amplified by the fact that the electrostatic pressure P_E is proportional to Φ^2, causing the instability. In contrast to the Rayleigh mode, the instability in the electrostatic mode is unavoidable at low wave numbers (i.e., long wavelength). This may explain why the filament always tends to branch into bushlike structures.

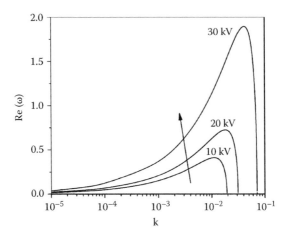

FIGURE 3.20
Instability growth rate ω at high applied voltages. k and ω are nondimensionalized as in Figure 3.19.

3.7 Nanosecond and Subnanosecond Discharge in Water

In the previous sections, the initiation and propagation of the electric breakdown process in liquid under microsecond high-voltage excitations were described. For the remaining question, what would happen if the high-voltage pulse duration is short enough so that no cavitation would have enough time to be formed? In this section, the formation of plasma discharge in water under nanosecond and subnanosecond pulse excitations is discussed, and the possibility to generate plasma in condensed media without density decrease is demonstrated.

3.7.1 Fast Imaging of Nanosecond and Subnanosecond Discharge in Water

Two different pulsed power systems were used. The first power supply generates pulses with +26-kV pulse amplitude in 50-ohm coaxial cable, 10-ns pulse duration, 0.3-ns rise time, and 3-ns fall time. The second system generates 112-kV pulses with 150-ps rise time, and duration on the half height about 500 ps. Typical examples of pulse shapes are presented in Figure 3.21 (Starikovskiy et al., 2011).

It was found that discharge in liquid water developed in a picosecond time scale. Size of the excited region near the tip of the high-voltage electrode was about 1 mm. The discharge had a complex multichannel structure (Figure 3.22), and this structure changed from pulse to pulse. Thus, an

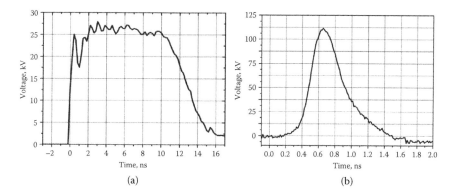

FIGURE 3.21
High-voltage pulses used in experiments. (a) nanosecond generator; (b) subnanosecond generator.

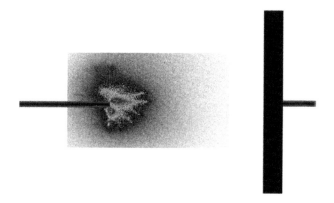

FIGURE 3.22
Geometry of discharge gap and ICCD camera field of view. Distilled water, $U = 27$ kV. Image was taken 5 ns after the pulse start. Camera gate was 1 ns; field of view was 2.6×1.7 mm.

image accumulation leads to discharge structure elimination but gives us the possibility to trace the emission dynamics with a 500-ps camera gate (Figure 3.23). Figure 3.23 demonstrates the dynamics of emission from the discharge taken with 1,000 accumulations. Image overlapping removed fine discharge structure but allowed better temporal resolution. It is clearly seen that the discharge emission has a typical duration close to the camera gate (500 ps), while the emission rise time has an even shorter time scale (~250 ps). The high-voltage rise time was about 150 ps, and we could assume that the typical discharge development time was close to these values (between 150 and 250 ps). The emitting region had a diameter of about 1 mm.

To analyze the spatial structure of the discharge, we used a longer camera gate (1 ns) without signal accumulation. As expected, the temporal resolution of the image decreased, and the observed emitting discharge phase became

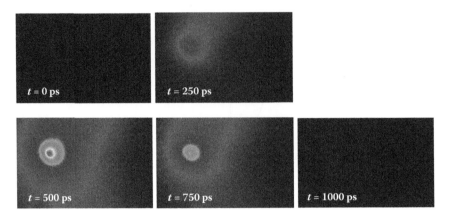

FIGURE 3.23
(See color insert.) Discharge emission dynamics. 1,000 accumulations/image. Camera gate was 500 ps, $\Delta\lambda$ = 250–750 nm. Image size was 7 × 4 mm.

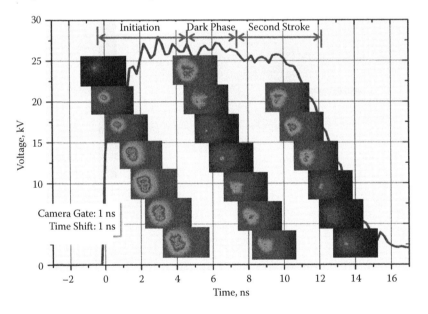

FIGURE 3.24
(See color insert.) Dynamics of discharge emission and high-voltage potential on electrode. Distilled water, U = 27 kV. Camera gate is 1 ns. Spectral response 250–750 nm.

longer (Figure 3.24). In exchange, we now had the spatial distribution of the emission, which looked like familiar gas-phase streamer discharge.

The diameter of the typical emitting channel was about 50 μm, where propagation length was 0.5–0.6 mm for U = 27 kV. In the case of a short rise time, we observed discharge propagation with a velocity of up to 2,000 km/s (2 mm/ns) during the initial stage of the discharge corresponding to the voltage rise time (Figure 3.24). When voltage reached the maximum, the discharge propagation stopped, and the "dark phase" appeared (t = 6–9 ns in Figure 3.24). During this phase, discharge could not propagate because of space charge formation and electric field decrease. Voltage decrease led to the second stroke formation and second emission phase (Figure 3.24, 10–13 ns). This means that the channels lost the conductivity, and the trailing edge of the nanosecond pulse generated a significant electric field and excitation of the media comparable to the excitation corresponding to the leading edge of the pulse.

No bubbling or void formation was observed during the present experiments. The discharge observed had a completely different nature from the discharges initiated by electrical pulses with a longer rise time (see, e.g., Malik et al., 2005; Lukes et al., 2008; An, Baumung, and Bluhm, 2007). In the publication by An, Baumung, and Bluhm (2007), the pulses with a 40-ns rise time at U = 18 kV demonstrated the velocity of discharge propagation about 2.5 km/s during the initial phase of the discharge and up to 35 km/s during the second phase. Both the shadow graph and Schlieren images suggest that the branches were gaseous.

The experiments at an elevated (up to 220 kV) voltage proved the discharge propagation in the liquid phase. The experiments were performed using ultrashort pulses with a typical duration of about 400 ps and a rise time of 150 ps (Figure 3.21b). Figure 3.25 demonstrates the integral image of the discharge (camera gate was 100 ns) for different voltages. It is clear that the discharge formation depended on the voltage applied. At 92 kV, discharge formation took longer than the pulse width, and only the very initial phase of the discharge formation could be observed (Figure 3.25a). Voltage increase elongated the plasma channels from 1–2 mm for 107 kV to 6–8 mm

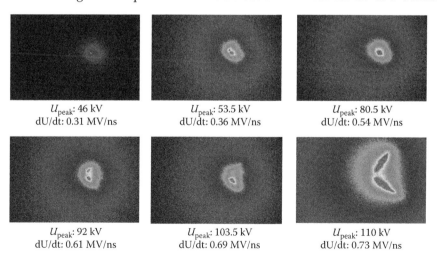

U_{peak}: 46 kV
dU/dt: 0.31 MV/ns

U_{peak}: 53.5 kV
dU/dt: 0.36 MV/ns

U_{peak}: 80.5 kV
dU/dt: 0.54 MV/ns

U_{peak}: 92 kV
dU/dt: 0.61 MV/ns

U_{peak}: 103.5 kV
dU/dt: 0.69 MV/ns

U_{peak}: 110 kV
dU/dt: 0.73 MV/ns

FIGURE 3.25
(See color insert.) Discharge development for different voltage. Pulse width was 400 ps. ICCD camera gate was 100 ns.

for 220 kV (Figures 3.25b–3.25f). Typically, we observed the formation of two or three channels from the top and bottom edges of the tip of the electrode. The length of the channels decreased gradually with the voltage decrease, which for a fixed pulse duration meant the streamer velocity decrease. For a voltage below 90 kV, discharge could not start for 400 ps.

As in the case of 10-ns pulses, no bubble formation was observed. Figure 3.26 shows the dynamics of the discharge formation for $U = 224$ kV. The Λ-shaped pulse had no plateau, and the dark phase of the discharge did not appear (Figure 3.26). Discharge development took an extremely short time (in 100–200 ps, plasma channels reached a length of 5–8 mm and a diameter of about 100 μm).

The observed propagation velocity reached 5 mm/ns (5,000 km/s; ~2% of speed of light in water) and was almost the same as a typical velocity of streamer propagation in air. Typical channel diameter is estimated as $d = 50$–100 μm. Thus, the radial expansion velocity of the channel could be estimated as $v_r = r/\Delta t \approx 0.05$ mm/200 ps = 250 km/s (Figure 3.27).

3.7.2 Ionization of Liquid by E-Impact

Next, we estimate the electrical field needed for water ionization by e-impact using the "dense gas" approach. We used cross sections from the BOLSIG+ Solver (Hagelaar, 2008) to analyze the ratio between the attachment rate of the electron and the rate of ionization (Figure 3.28).

Figure 3.29 demonstrates the rate coefficients for different processes in low-density water vapor. Since these data cannot give a quantitative description of the processes in liquid water, we use them for evaluation purposes only.

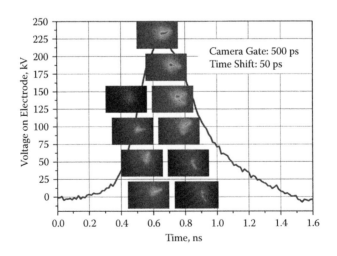

FIGURE 3.26
(See color insert.) Dynamics of discharge emission. Distilled water. $U = 220$ kV. $dU/dt = 1.46$ MV/ns. Camera gate was 500 ps. Time shift between frames was 50 ps.

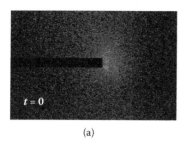 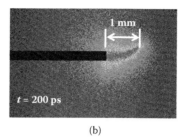

(a)   (b)

FIGURE 3.27
(See color insert.) Velocity of streamer propagation in water. 150 ps rise time, 1.5 MV/ns: Growth rate $v_z = L/\Delta t \approx 1$ mm/200 ps = 5,000 km/s ~ 15% c. Expansion rate $v_r = r/\Delta t \approx 0.05$ mm/200 ps = 250 km/s.

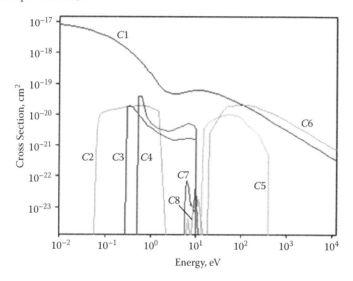

FIGURE 3.28
(See color insert.) Excitation cross section (m²) for H_2O vapor (rarefied gas). C_1, elastic collisions; C_2, rotational excitation (0.02 eV); C_3, vibrational excitation (0.20 eV); C_4, vibrational excitation (0.45 eV); C_5, electronic excitation (7.10 eV); C_6, ionization (12.61 eV); C_7, C_8, dissociative attachment.

From Figure 3.29, it is clearly seen that the ionization rate becomes higher than the attachment for $E/n > 50$ Td. In condensed media, multibody attachments will play the major role. Taking into account a strong dependence of the rate constant of ionization from E/n, we can assume, nevertheless, that for $E/n > 100$ Td we can obtain fast ionization even in condensed media. At normal conditions, it means that the electric field value should be close to or greater than $E_{crit} > 30$ MV/cm. Almost the same estimation ($E \sim 10$–20 MV/cm) was obtained by Kolb et al. (2008).

Taking into account the geometrical parameters of the plasma channels (Figure 3.27), we could estimate the radius of the tip of the streamer as

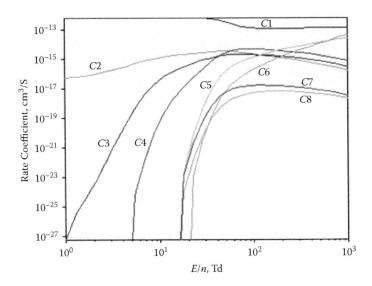

FIGURE 3.29
(See color insert.) Rate coefficients (m³/s) for different processes in water vapor in dependence on the E/n value. C_1, transport collisions; C_2, rotational excitation; C_3, C_4, vibrational excitation; C_5, electronic excitation; C_6, ionization; C_7, C_8, dissociative attachment.

$r \sim 10$ mm and subsequent electric field as $E \sim 220$ MV/cm ($E/n \sim 670$ Td). Thus, the e-field near the discharge tip was an order of magnitude higher than the critical electric field estimated using the dense gas model. Thus, we can assume that the direct ionization by electron impact takes place in the streamer head even in liquid water, and it is possible to generate liquid nonequilibrium plasma in different media.

3.7.3 Chance of Voids Formation

Water in laboratory conditions always contains trace impurities of dissolved gases. There are several different mechanisms for formation of vapor bubbles (voids) in the liquid under discharge conditions (Bruggeman and Leys, 2009): electrostatic mechanism, evaporation due to conductive current and joule heat release, and initial bubble expansion. In our experiments, we were able to control the bubbles larger than 5 μm and did not find any trace of them before and after the discharge. Below this limit, which was determined by the spatial resolution of the photocathode and optical interference, one can assume existence of the bubbles. Nevertheless, the size of these initial bubbles should be very small in comparison with the plasma channel diameter (~100 μm). Thus, we can analyze the possibility of an initial bubble expansion from possible "detection-limited size" to "plasma channel size" during the discharge development time (~100 ps). It is clear that to form the void in the liquid we need to generate a strong shock wave to accelerate the liquid from the axis of the discharge channel. Simple estimations gave us the required flow velocities of about 100–200 km/s in the radial direction. The question is whether we achieve these parameters using low-energy discharge in water.

The analysis showed that the liquid motion with a velocity 30–35 km/s (with a compression ratio of 3.8) is possible only under nuclear test explosion experiments (Trynin, 1994; Avrorin et al., 1980; Avrorin et al., 2006). One of the tests (Trynin, 1994) demonstrated a shock wave propagation velocity of $D = 43.95$ km/s, velocity of water behind the shock wave front $U = 32.54$ km/s, compression ratio $s = r/r_0 = 3.852$, and pressure behind shock wave front $P = 1,430$ GPa at a distance of 2 m from the epicenter of the underground nuclear explosion. Total discharge energy in the present experiments allowed us to overheat the water in the discharge channel for 50 K only and definitely not enough to generate an ultrastrong shock wave to create the void.

Thus, pulsed discharges in the subnanosecond range allowed us to generate highly ionized channels in condensed media without void formation. This approach creates numerous possibilities for "liquid plasma" applications, like material synthesis, biophysical and biomedical plasma systems, light sources, and liquid-phase plasma-chemical systems.

4

Decontamination of Volatile Organic Compounds

4.1 Introduction

Volatile organic compounds (VOCs) are carbon-containing organic chemicals that have low water solubilities and high vapor pressures. Under normal conditions, they significantly vaporize and enter the atmosphere. Due to their high volatility at low temperatures, VOCs are mobile and therefore more likely to be released to the environment than ordinary organic contaminants.

VOCs form a class of water pollutant that has been addressed by environmental regulation in the past several decades due to the toxicity and contribution to global warming mechanisms of the toxicants. The U.S. Environmental Protection Agency (EPA) estimated that VOCs are present in one-fifth of the nation's water supplies. The control of water pollution from large-volume sources, such as paper mills, pharmaceuticals, and food-processing plants, has become a challenging problem. The most common VOCs that are found in drinking water include chlorinated solvents, such as carbon tetrachloride, 1,2-dichloroethane, 1,1-dichloroethylene, *trans*-1,2-dichloroethylene, methylene chloride, and tetrachloroethylene, and fuel components such as benzene, methyl tert-butyl ether (MTBE), toluene, and xylenes. They enter groundwater from a variety of sources. Benzene, for example, a fundamental component of gasoline and diesel fuel, may enter groundwater from gasoline or oil spills on the ground surface or from leaking underground fuel tanks, while some chlorinated solvents are found in such household products as spot removers, typing correction fluids, adhesives, automotive cleaners, inks, and wood furniture cleaners. A number of them (e.g., MTBE and chlorofluorocarbons [CFCs]) have been banned in the United States to limit further contamination of drinking water aquifers. Under the Safe Drinking Water Act (SDWA), the EPA has promulgated maximum contaminant levels (MCLs) for 83 specific drinking water contaminants, including 23 VOCs as summarized in Table 4.1.

TABLE 4.1

EPA Regulation of Volatile Organic Chemicals

VOC	Promulgation Date	MCLG (mg/L)	MCL (mg/L)	Carcinogen Ranking
Benzene	7/87	0	0.005	A
Carbon tetrachloride	7/87	0	0.005	B2
1,2-Dichloroethane	7/87	0	0.005	B2
1,1-Dichloroethylene	7/87	0.007	0.007	C
Para-dichlorobenzene	7/87	0.075	0.075	C
1,1,1-Trichloroethane	7/87	0.20	0.20	D
Trichloroethylene	7/87	0	0.005	B2
Vinyl chloride	7/87	0	0.002	A
Dibromochloropropane	12/90	0	0.0002	B2
o-Dichlorobenzene	12/90	0.6	0.6	D
cis-1,2-Dichloroethylene	12/90	0.07	0.07	D
trans-1,2-Dichloroethylene	12/90	0.1	0.1	D
1,2-Dichloropropane	12/90	0	0.005	B2
Ethylbenzene	12/90	0.7	0.7	D
Ethylene dibromide	12/90	0	0.00005	B2
Methylene chloride	12/91	—	—	B2
Monochlorobenzene	12/90	0.1	0.1	D
Styrene	12/90	0.1	0.1	B2
Tetrachloroethylene	12/90	0	0.005	B2
Toluene	12/90	2	2	D
Trichlorobenzene	12/91	—	—	D
1,1,2-Trichloroethane	12/91	—	—	C
Xylene	12/90	10	10	D

Source: Ram, N.M., Christman, R.F., and Cantor, K.P. (1990) *Significance and treatment of volatile organic compounds in water supplies,* CRC Press, Chelsea, MI.

Note: MCL, maximum contaminant level; MCLG, maximum contaminant level goal. *EPA carcinogen classification:* Group A, human carcinogen based on sufficient evidence from epidemiological studies; group B1, probable human carcinogen based on at least limited evidence of carcinogenicity to humans; group B2, probable human carcinogen based on a combination of sufficient evidence in animals and inadequate data in humans; group C, possible human carcinogen based on limited evidence of carcinogenicity in animals in the absence of human data; group D, not classified based on inadequate evidence of carcinogenicity from animal data; group E, no evidence of carcinogenicity for humans.

4.2 Conventional Technologies

Chlorination is the most widespread technology for water decontamination. Chlorine gas and its compounds (calcium or sodium hypochlorite, for example) function mostly as an oxidative agent during the decontamination process. However, chlorine is not suitable for removal of VOCs because of the possibility of the production of chlorinated organic compounds, including dioxins or dioxin-like compounds, which have significant toxicity to plants, animals, and humans.

In the water treatment industry, air stripping is a process to transfer volatile components in a liquid into an air stream and commonly is used for the removal of VOCs, including BTEX (benzene, toluene, ethylbenzene, and xylene) compounds found in gasoline, and solvents, including trichloroethylene and tetrachloroethylene from different contaminated water sources due to their relatively high vapor pressures. Reverse osmosis and active carbon filtration are other frequently used approaches for removing organic molecules from industrial effluents. However, the cost of these technologies is high, and they are generally not economical for large flow rates and low VOC concentrations. These physical processes are not actually destroying the hazardous contaminants, and the air or water effluents exiting the system may still require additional emission control.

In the past several decades, several advanced oxidation technologies (AOTs), including ozone oxidation, electrochemical oxidation, ultraviolet radiation photolysis, and hydrogen peroxide oxidation, have been studied for the degradation of organic pollutants in wastewater. The effectiveness of AOTs relies on the generation of reactive species from various pathways, particularly hydroxyl radicals to oxidize and ultimately destroy organic pollutants. Among the emerging low-temperature VOC treatment technologies are low-temperature catalysis, biofiltration, and nonthermal plasma. Drawbacks of the catalysts are plugging, fouling, or poisoning by particles and non-VOC materials in the exhaust stream, resulting in high maintenance cost. The major disadvantage of the biofilter is its large specific footprint. Biofilter systems and filter materials may also require costly maintenance and replacement.

The application of nonthermal plasma directly in or above water has a potential to solve most of the problems typical for alternative VOC treatment methods. The application of nonthermal plasma can decrease energy costs by the direct injection of electric energy into aqueous solutions, where different reactive chemical radicals and molecular species, including OH, H, O, HO_2, and H_2O_2, as well as physical processes such as ultraviolet radiation and shock waves can be formed. Plasma approaches usually are more energy efficient and competitive than other cleaning methods when stream volumes are large and the concentration of pollutants is small. Atmospheric pressure plasma systems that have been investigated for VOC abatement include pulsed corona, pulsed spark, dielectric barrier discharge (DBD), gliding arc discharge, and liquid-gas hybrid discharges. Nonthermal plasma systems have been successfully applied for cleaning of high-volume, low-concentration (HVLC) VOC exhausts, particularly from volatile hydrocarbons (methanol, ethanol, acetone, etc.); aromatic compounds (benzene, phenol, toluene, etc.); chlorine-containing compounds (trichloroethylene, trichloroethane, carbon tetrachloride, etc.); organic dyes (rhodamine B, methyl orange, methyl blue, Chicago sky blue, etc.); and pharmaceutical compounds (pentoxifylline, carbamazepine, clofibric acid, and iopromide, etc.). Some of the processes are specifically discussed in the next section.

4.3 Mechanism of Plasma Treatment of VOCs

Electric energy cost is one of the critical factors characterizing the plasma treatment of VOCs. To analyze the energy cost of VOC treatment in nonthermal plasma, it should be noted that the mechanism of plasma oxidation of hydrocarbons can be generally interpreted as the low-temperature burning of the VOCs in water to preferentially CO_2 and H_2O using plasma-generated active oxidizers, including OH, atomic oxygen, electronically excited oxygen, ozone, and so on. Note that OH radicals are of special importance for the oxidation process (Rosario-Ortiz et al., 2008).

The nonthermal plasma oxidation of low-concentration volatile hydrocarbons in water can be generally outlined by the illustrative kinetic mechanism that follows. Among different plasma-generated active oxidizers (OH radicals, atomic oxygen, electronically excited oxygen, ozone, etc.) considered for the cold burning of VOCs, OH radicals play a major role. The formation of OH radicals in underwater plasma is due to the presence of water molecules and is provided through numerous channels. However, it is interesting to point out a selective one, starting with the charge exchange from any positive ions M^+ to water ions, H_2O^+ as described by the following equation:

$$M^+ + H_2O \rightarrow M + H_2O^+ \qquad (4.1)$$

The H_2O^+ ions, formed through the charge exchange between M^+ and H_2O molecules, react with water molecules to produce active OH radicals in the fast ion-molecular reaction:

$$H_2O + H_2O^+ \rightarrow H_3O^+ + OH \qquad (4.2)$$

Taking into account that the ionization potential of the water molecules is relatively low (see Table 4.2 for properties of the H_2O molecule in different phases), most of the positive ions initially formed in the plasma channel have a tendency for charge exchange. Therefore, the discharge energy initially distributed over different components is somewhat selectively localized on the ionization of water molecules and the selective production of OH radicals. It explains the relatively low energy price of the production of OH radicals in air plasma, which is about 10–30 eV/radical. If we represent a hydrocarbon VOC molecule as RH, where R is a relevant organic group, the oxidation of the molecule starts with the following elementary reaction of its dehydrogenization. (Choi et al., 2006; Billamboz et al., 2010):

TABLE 4.2

Properties of H_2O Molecule in Different Phases

Phase	O-H Bond Length (Å)	O-O Bond Distance (Å)	Ionization Potential (eV)
Gas	0.96	∞ (free)	12.61
Liquid	0.96	∞ (free)	10.56
Ice	1.01	2.76	11.00

$$RH + OH \rightarrow R + H_2O \tag{4.3}$$

Almost immediate attachment of molecular oxygen to the active organic radical R results in the formation of an organic peroxide radical:

$$R + O_2 \rightarrow RO_2 \tag{4.4}$$

The peroxide radical RO_2 is able to react with another saturated hydrocarbon VOC molecule RH, forming saturated organic peroxide RO_2H and propagating the chain reactions in Equations 4.3 and 4.4 of RH oxidation as described by the following equation:

$$RH + RO_2 \rightarrow RO_2H + R \tag{4.5}$$

Depending on the temperature and chemical composition, peroxide RO_2 and RO_2H are further oxidized to CO_2 and H_2O with or without additional consumption of OH and other plasma-generated oxidizers. Summarizing the mechanism, the energy cost corresponding to the VOC treatment process is on the level of 10–30 eV per molecule of the pollutant RH. This energy is relatively high, but the total energy consumption can be small because of the very low concentration of pollutants (usually on the order of parts per million). A major conventional approach to VOC control, regenerative thermal oxidizers (RTOs), consumes about 0.1 eV per molecule, but this energy is calculated per molecule of water. Therefore, the energy consumption in plasma becomes lower than that of the conventional RTO when the VOC concentration in water is low.

4.4 Decomposition of Methanol and Ethanol

Generally, low concentrations of methanol (CH_3OH) and ethanol (C_2H_5OH) are not considered as significant sources of organic compound contamination in water solutions, although at higher doses methanol may cause permanent blindness by the destruction of optical nerves in the eye. Still, the plasma-assisted decomposition of low-chain alcohols like methanol and ethanol lets them be used as the model substances of water-soluble organic wastes based on the following facts: First, the structures of methanol and ethanol are relatively simple, and the study of their decomposition mechanism may help the understanding of more complicated hydrocarbons. Second, the conversion of these low-chain alcohols is an attractive process for obtaining hydrogen or syngas. Both alcohols are considered excellent liquid H_2 sources and are attainable from the conversion of fossil fuels or natural sources. Their use, especially in the case of methanol, is mainly intended for hydrogen production for fuel cells and automotive applications. Hydrogen can be easily stored and handled in the form of liquid alcohol, thereby avoiding the use of high-pressure conditions, risk of leakage and explosions, and exhaustive storage and safety protocols.

The decomposition mechanisms of methanol and ethanol by plasma atmospheric pressure plasma were studied by several groups (Yukhymenko et al., 2008; Kostyuk, 2008; Nishioka, Saito, and Watanabe, 2009; Yan, Li, and Lin, 2009; Rico et al., 2010; Derakhshesh, Abedi, and Hassanzadeh, 2010). A DC (direct current) water plasma torch with power of 0.91–1.05 kW and a current of 7.0 A was used, through which methanol and ethanol solutions with mole fractions ranging from 1% to 100% were injected, with a feed rate of plasma-supporting gas produced from the solution ranging from 40 to 80 mg/s (Nishioka, Saito, and Watanabe, 2009). The decomposition rates of ethanol and methanol as a function of molar concentration are shown in Figure 4.1. The decomposition rates strongly depended on the molar concentration, and the plasma technology was especially efficient at lower concentrations (£8%). By-products such as CH_2, CH_3, CH_4, C_2H_2, CO, H_2, and so on were detected. Figure 4.2 summarizes the main path of methanol and ethanol decomposition in the thermal DC water plasma torch (Nishioka, Saito, and Watanabe, 2009).

For hydrogen production from methanol and ethanol, nonthermal plasmas are usually utilized to avoid the pyrolysis of organic molecules. Taking methanol, for example, DBD and atmospheric pressure glow discharge are frequently used for the dissociation (Yan, Li, and Lin, 2009; Rico et al., 2010). According to previous studies, there are several initial reaction channels of hydrogen production. Among them is the dissociation of methanol molecules due to the relative lower dissociation energy:

$$e + CH_3OH \rightarrow CH_3O + H + e \qquad (4.6)$$

$$e + CH_3OH \rightarrow CH_3 + OH + e \qquad (4.7)$$

$$e + CH_3OH \rightarrow CH_2OH + H + e \qquad (4.8)$$

$$e + CH_3OH \rightarrow CH_2 + H_2O + e \qquad (4.9)$$

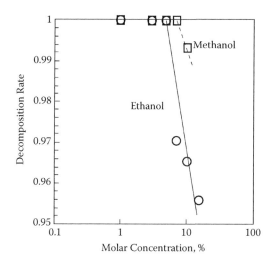

FIGURE 4.1
Decomposition rates of methanol and ethanol solutions in DC water plasma torch. (From Nishioka, H., Saito, H., and Watanabe, T. (2009) Decomposition mechanism of organic compounds by DC water plasmas at atmospheric pressure. *Thin Solid Films* 518, 924–928.)

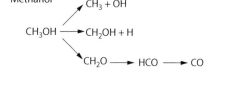

FIGURE 4.2
Decomposition mechanism of methanol and ethanol in thermal DC water plasma torch. (From Nishioka, H., Saito, H., and Watanabe, T. (2009) Decomposition mechanism of organic compounds by DC water plasmas at atmospheric pressure. *Thin Solid Films* 518, 924–928.)

$$CH_3OH \rightarrow HCHO + H_2 \qquad (4.10)$$

The initial reactions of dissociation are induced by high-energy electrons with the plasma discharge, directly generating radicals, including CH_2, CH_3, CH_3O, CH_2OH and H. Reaction 4.10 directly generates H_2, while the other reactions generate H, which then recombines with each other to produce hydrogen.

4.5 Decomposition of Aromatic Compounds

Aromatic compounds, including benzene, toluene, phenol, and so on, make up a big part of organic contaminants in water. Phenol, for instance, is produced on a large scale (about 7 billion kilograms/year) as a precursor to many materials and useful compounds. The major uses of phenol involve its conversion to plastics or related materials. Because phenol is used in many manufacturing processes and in many products, exposure to phenol may take place at work or at home in very low amounts. Phenol is present in a number of consumer products that are swallowed, rubbed on, or added to various parts of the human body. These include ointments, ear and nose drops, cold sore lotions, mouthwashes, toothache drops, analgesic rubs, throat lozenges, and antiseptic lotions. Phenol has been found in drinking water, automobile exhaust, tobacco smoke, and certain foods, including fried chicken and some species of fish.

Different plasma discharges, including underwater corona discharge, diaphragm discharge, gliding arc discharge above water, and hybrid gas-liquid discharge, are used for the removal of phenol in water solution, and effective decomposition has been demonstrated in these plasma systems (Sun, Sato, and Clements, 1999; Sugiarto and Sato, 2001; Grymonpre et al., 2001; Manolache Shamamian, and Denes, 2004; Chen et al., 2004; Kusic, Koprivanac, and Locke, 2005; Lukes and Locke, 2005; Liu and Jiang, 2005; Yan et al., 2005; Shin et al., 2000; Grabowski et al., 2006; Li, Zhou, et al., 2007; Dors, Metel, and Mizeraczyk, 2007; Shen et al., 2008; Yang, Zhang, et al., 2009). The removal

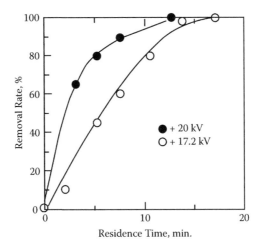

FIGURE 4.3
Phenol removal rate by underwater corona discharge as a function of residence time. (Sun, B., Sato, M., and Clements, J.S. (1999) Use of a pulsed high-voltage discharge for removal of organic compounds in aqueous solution. *J. Phys. D: Appl. Phys.* 32, 1908–1915.)

rate of phenol is presented in Figure 4.3 as a function of treatment time (i.e., input plasma energy), which was obtained utilizing the underwater corona discharge system developed by Sun, Sato, and Clements (1999). The final products of nonthermal plasma treatment are CO_2 and H_2O. At lower initial concentrations, no phenol has been detected in aftertreatment solutions. Several intermediate products were detected during the treatment, including 1.4-benzoquinone, pyrocatechol, and hydroquinone. It is generally believed that OH radicals play an important role in the plasma-assisted decomposition of phenol, and the mechanism of plasma removal of phenol from water solution starts with oxidation provided by OH radicals:

$$C_6H_5OH + OH \rightarrow C_6H_5(OH)_2 \tag{4.11}$$

The dihydroxylcyclohexadienyl radical $C_6H_5(OH)_2$ can decay to form the phenoxyl radical C_6H_5O, which further reacts with OH radical to form hydroquinone and pyrocatechol:

$$C_6H_5O + OH \rightarrow C_6H_4(OH)_2 \tag{4.12}$$

The intermediate products formed will be further oxidized under the presence of plasma-generated oxidizers and oxygen, resulting in the opening of the aromatic ring and formation of low molecular weight hydrocarbons.

The effects of different chemical additives on phenol degradation were studied by different research groups (Grymonpre et al., 2001; Chen et al., 2004; Kusic, Koprivanac, and Locke, 2005; Liu and Jiang, 2005). Hydrogen peroxide was added to the plasma reactor, and the degradation rates of phenol increased even though the direct reaction of phenol with hydrogen peroxide was very slow at the ambient temperature. The degradation process was believed to be due to the reaction of hydroxyl radicals formed by the photolysis of hydrogen peroxide, caused by the ultraviolet radiation from electrical discharge plasma:

$$H_2O_2 + h\upsilon \rightarrow 2OH \tag{4.13}$$

Fenton's reagent has long been used to treat a large variety of water pollutions. The addition of ferrous sulfate solution in the plasma reactor has provided a much higher removal rate in comparison to the non-ferrous-sulfate solutions (Grymonpre et al., 2001). The effect of ferrous salt can be attributed to two factors. First is the traditional Fenton's reaction, in which ferrous iron is oxidized to ferric iron, producing hydroxyl radical:

$$Fe^{2+} + H_2O_2 \rightarrow Fe^{3+} + OH + OH^- \tag{4.14}$$

The second is the direct oxidation of phenol decomposition intermediates from Equation 4.12 by Fe^{3+}:

$$Fe^{3+} + C_6H_4(OH)_2 \rightarrow Fe^{2+} + C_6H_4O_2H + H^+ \tag{4.15}$$

TABLE 4.3

Energy Efficiency for Phenol Removal in a Hybrid Gas-Liquid Plasma Reactor

		Removal (%)	g/kWh	kWh/m³ per Log Removal
Without ferrous sulfate	No carbon, stagnant air	24.6	0.3	791
	No carbon, O_2 flowing (150 sccm)	35.6	0.7	316
	Carbon (1 g/L), O_2 flowing (150 sccm)	100	3.7	30
With ferrous sulfate	No carbon, stagnant air	89.8	1.4	107
	No carbon, O_2 flowing (150 sccm)	97.5	2.2	56
	Carbon (1 g/L), O_2 flowing (150 sccm)	100	4.0	22

Source: Grymonpre et al., 2001.
Note: sccm, standard cubic centimeter per minute.

$$Fe^{3+} + C_6H_4O_2H \rightarrow Fe^{2+} + C_6H_4O_2 + H^+ \qquad (4.16)$$

In addition to hydrogen peroxide and ferrous sulfate, oxygen gas and activated carbon were added in the plasma reactor to further improve the energy efficiency. Table 4.3 presents the energy consumption in a hybrid gas-liquid plasma reactor for phenol removal.

In addition to phenol, other aromatic compounds, such as benzene, toluene, cresol, and so on, have been shown to be oxidatively degraded toward inorganic end products with different plasma treatment methods either directly in solution or suspended above the liquid surface (Ogata et al., 2000; Einaga, Ibusuki, and Futamura, 2001; Manolache, Shamamian, and Denes, et al., 2004; Tomizawa and Tezuka, 2006; Liu et al., 2007; Schmid, Jecklin, and Zenobi, 2010).

4.6 Decomposition of Chlorine-Containing Compounds

Chlorine-containing compounds are widely used in the industrial, agricultural, and medical fields. Plasma systems have been effectively applied in the destruction of different chlorine-containing VOCs, including vinyl chloride, trichloroethylene, trichloroethane, and carbon tetrachloride (Kirkpatrick, Finney, and Locke, 2003; Marotta, Scorrano, and Paradisi, 2005; Huang et al., 2007; Baroch et al., 2008; Trevino and Fisher, 2009). Considering as an example the destruction of carbon tetrachloride (CCl_4) diluted in water, a major contribution to the process kinetics is provided by plasma-generated OH radicals, O and N atoms, as well as direct electron-impact destruction through dissociative attachment (Penetrante et al., 1996). The radical-induced decomposition starts with the formation of OH radicals and O and N atoms:

$$e + O_2 \rightarrow O(^3P) + O(^1D) + e \tag{4.17}$$

$$O(^1D) + H_2O \rightarrow 2OH \tag{4.18}$$

$$e + N_2 \rightarrow N(^4S) + N(^2D) + e \tag{4.19}$$

The oxidation of CCl_4 is initiated by its reactions with O and OH:

$$OH + CCl_4 \rightarrow HOCl + CCl_3 \tag{4.20}$$

$$O(^3P) + CCl_4 \rightarrow ClO + CCl_3 \tag{4.21}$$

The major initial reaction of CCl_4 decomposition through dissociative attachment starts with the following elementary process:

$$e + CCl_4 \rightarrow Cl^- + CCl_3 \tag{4.22}$$

The major products of plasma processing of CCl_4 are Cl_2, $COCl_2$, and HCl. These products can be easily removed as they dissolve or dissociate in aqueous solutions and combine with $NaHCO_3$ in a scrubber solution to form NaCl. Similar approaches can be applied as performed for product removal during the plasma cleaning of other diluted chlorine-containing VOC exhausts.

The plasma-stimulated decomposition of CCl_4 is strongly dependent on the dissociative attachment, consuming an electron per each decomposition. Therefore, the energy cost of the CCl_4 decomposition is determined by the energy cost of ionization. The ionization energy cost is lower at higher electron energies in a plasma system, and the lowest ionization energy cost can be achieved in plasma generated by electron beams. This explains the high energy efficiency of electron beams in the destruction of carbon tetrachloride, as shown in Figure 4.4.

Another frequently studied model chlorine-containing compound is chlorophenol. It was found that hydroxylation of aromatic rings proceeded as a key step in the initial stage of the degradation process, and it was confirmed that most of the chlorine atoms present in the reactants consumed were liberated as chloride ions. Tezuka and Iwasaki (1998) summarized the reaction paths of three isomeric monochlorophenols as shown in Figure 4.5. The attack of the hydroxyl radicals occurred most favorably at the para position to the OH group, less favorably at the ortho position, and very little at the meta position due to the electron-donating character of the phenolic OH group and the electrophilicity of the radicals. As a result, the difficulty of dechlorination increased in the order of para-, ortho-, and meta-chlorophenol.

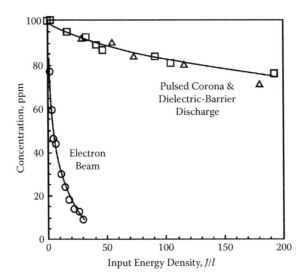

FIGURE 4.4

Electron beam, pulsed corona, and DBD treatment of carbon tetrachloride. (From Penetrante, B.M., Hsiao, M.C., Bardsley, J.N., Merritt, B.T., Vogtlin, G.E., and Wallman, P.H. (1996) Electron beam and pulsed corona processing of volatile organic compounds in gas streams. *Pure Appl. Chem.* 68, 1083–1087.)

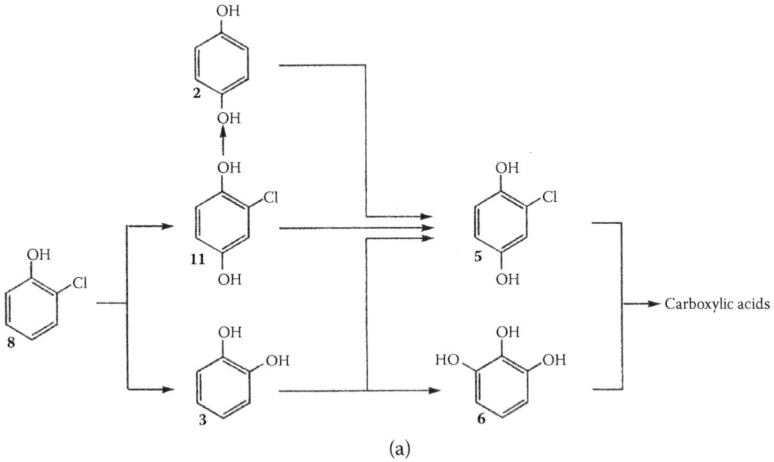

(a)

FIGURE 4.5

Reaction paths of ortho-, meta-, and para-chlorophenol. (From Tezuka, M., and Iwasaki, M. (1998) Plasma induced degradation of chlorophenols in an aqueous solution. *Thin Solid Films* 316, 123–127.)

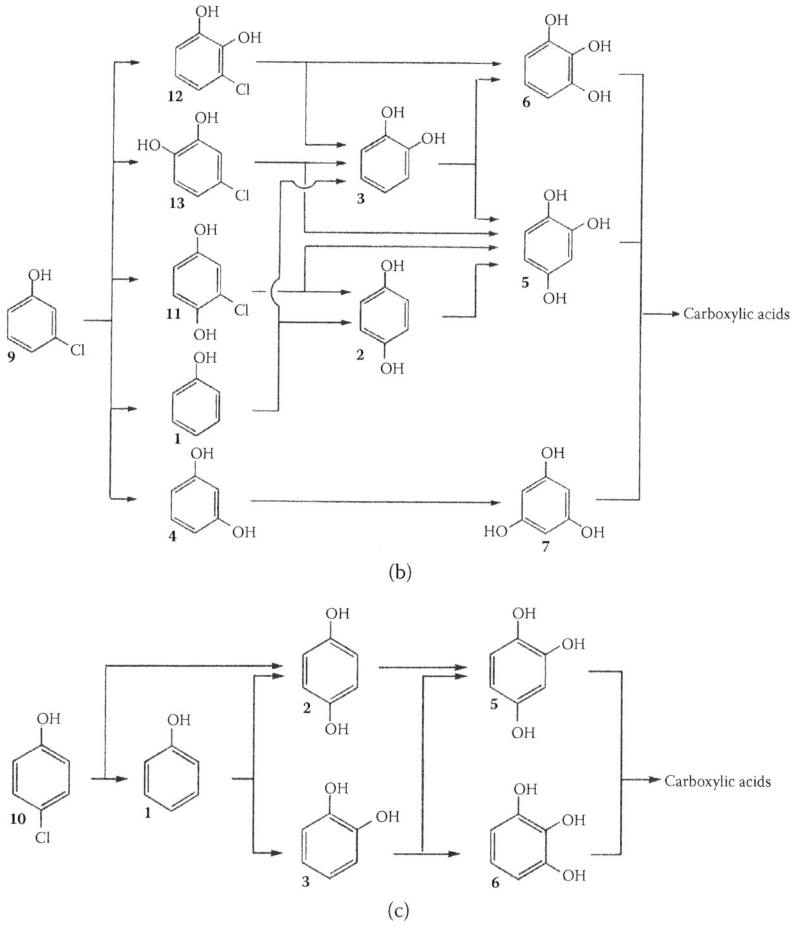

(b)

(c)

FIGURE 4.5 (CONTINUED)
Reaction paths of ortho-, meta-, and para-chlorophenol. (From Tezuka, M., and Iwasaki, M. (1998) Plasma induced degradation of chlorophenols in an aqueous solution. *Thin Solid Films* 316, 123–127.)

4.7 Decoloration of Dyes in Wastewater

A variety of new synthetic organic dyes is frequently used by the modern textile and food industries. Although the concentration of these dyes in effluent is usually low, they are easily visible in wastewater. Due to differences in the chemical composition of dyes, there is no universally applicable chemical technology for the removal of dyes from wastewater. Therefore, the removal of these dyes from effluent by a nonselective method to simplify the decoloration process has become a major environmental challenge, not only because of the potential toxicity of certain dyes, but also due to their visibility in water and subsequent heightened concerns from the public.

The decoloration of different dye solutions, including rhodamine B, methyl orange, methyl blue, Chicago sky blue, acridine orange, a-naphthol, red B, flavine G, acid orange 7, indigo carmine, and more, have been studied by different plasma discharges (Sugiarto, Ohshima, and Sato, 2002; Sugiarto et al., 2003; Gao et al., 2003; Njatawidjaja et al., 2005; Ghezzar et al., 2007; Ishijima et al., 2007; Magureanu et al., 2008; Nicolae, 2008; Li, Zhou, et al., 2007; Zhang et al., 2007; Minamitani et al., 2008; Wang et al., 2008; Handa and Minamitani, 2009; Sung et al., 2010). The breaking up of the bonds between chromophores by oxidative radicals is considered to be the dominant mechanism of dye decoloration. The decoloration rate is strongly dependent on the initial dye concentration. Figure 4.6 shows an example of decrease in the rate of decoloration for rhodamine B at higher concentrations.

The decoloration process is usually believed to be the combination of several effects. In addition to the reactive radicals (especially OH) generated, one of the physical effects produced by pulsed discharge plasma in water is ultraviolet radiation. A direct photooxidation of dye with ultraviolet light alone can lead to the decoloration of the dye through the following reaction:

$$RH + h\nu \rightarrow R^* + H \tag{4.23}$$

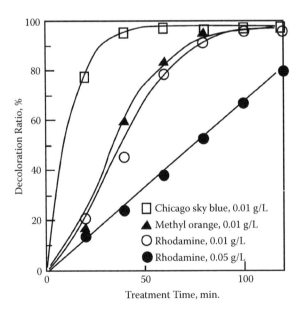

FIGURE 4.6
Decoloration of aqueous dye solutions by pulsed corona discharge treatment. (From Sugiarto, A.T., Ito, S., Ohshima, T., Sato, M. and Skalny, J.D. (2003) Oxidative decoloration of dyes by pulsed discharge plasma in water. *J. Electrostat.* 2003, 58, 135–145.).

An increase of decoloration efficiency was observed when the underwater plasma was switched from corona mode to spark mode. The intensity of ultraviolet radiation in the case of spark discharge was much higher than that of the corona discharge mode (Sugiarto et al., 2003), indicating the direct correlation between the decoloration process and the ultraviolet radiation. However, direct photooxidation in water was not very energy efficient since ultraviolet radiation with a wavelength below 200 nm was strongly absorbed by water. Therefore, ultraviolet radiation is usually used in combination with oxidants (e.g., hydrogen peroxide) or catalysts (e.g., TiO_2) to form hydroxyl radicals. The hydroxyl radicals can further react with dye through Reaction 4.3 and lead to decoloration. In these cases, the production of hydroxyl radicals by electron impact dissociation, which usually dominates in corona or spark mode, becomes less important.

It is also well known that the oxidation processes are sensitive to the pH of the aqueous solutions. It was reported that the decoloration of dyes using a photolysis process was more effective at low pH values than at high pH values (Perkins, 1999). Kang et al. (2000) reported that the optimum pH for both the formation of hydroxyl radicals and dye removal in a photo-Fenton process ranged from 3.0 to 5.0.

4.8 Decomposition of Freons (Chlorofluorocarbons)

Thermal and transitional (quasithermal) plasma systems can be effectively applied for decomposition, incineration, and re-forming of concentrated industrial and municipal wastes. The decomposition of freons (CFCs) can be considered an example of such waste treatment processes. Freons have been widely used in cooling systems, in the production of insulation foams and aerosols, and as solvents, especially in the electronic industry. The total amount of freons industrially produced is large, and they can penetrate to the upper atmosphere, destroying the ozone layer. Solar ultraviolet radiation dissociates CFC and produces Cl atoms, which stimulate a chain reaction of ozone destruction:

$$Cl + O_3 \rightarrow ClO + O_2 \tag{4.24}$$

$$ClO + O \rightarrow Cl + O_2 \tag{4.25}$$

The process repeats itself until the Cl is rained out as HCl, and a single Cl atom is able to destroy up to 10^6 ozone molecules. For these reasons, CFC production has been prohibited by the Montreal Protocol since 1987. Any remaining stocks of freons by now have been destroyed. Decomposition of the freons in thermal plasma jets (Murphy, 1997) and under direct action of electric discharges (Kohchi, Adachi, and Nakagawa, 1996; Nakagawa, Adachi and Kohchi, 1996; Wang et al., 1999; Jasiński et al., 2002) is an effective practical approach to CFC destruction.

Most of plasma CFC destruction processes are carried out in the presence of oxygen (Wang et al., 1999); air and water (Kohchi, Adachi, and Nakagawa, 1996; Jasiński et al., 2002); hydrogen and oxygen (Nakagawa, Adachi, and Kohchi, 1996); hydrogen (Wang et al., 1999); and nitrogen (Jasiński et al., 2002). An Australian company (SRL Plasma Ltd.) decomposes halocarbon compounds with a destruction degree of 99.99% in argon plasma jets in the presence of oxygen and water vapor or water vapor alone (Murphy, 1997; Murphy et al., 2002). The major decomposition products under such conditions are CO_2, HCl, and HF. The presence of water vapor significantly reduces the production of other unwanted products, such as other freons, dioxins, and furans. Although CFC destruction effectiveness in thermal plasma jets is excellent, this technology is characterized by significant heat losses and relatively high electric energy cost. The electric energy cost of plasma CFC destruction can be reduced by performing the process not in a thermal plasma jet but directly in the active discharge zone. Direct decomposition of freons in plasma has been organized at atmospheric pressure as well as at reduced pressure. The concentration of freons varied from very low (several dozen parts per million) in diluted mixtures to very high (several dozen percent) in nondiluted exhausts.

Effective decomposition of $CFCl_3$, CF_2Cl_2, and CHF_2Cl was achieved in the transitional regimes of gliding arc discharges in the presence of water vapor (Czernichowski, 1994) and oxygen (Opalska, Opalinska, and Ochman, 2002). An admixture of H_2 to freons during their treatment in gliding arc discharges limits products to simple aliphatic hydrocarbons, completely avoiding the formation of chlorinated dioxins and furans or phosgene, which can be generated in oxidative or redox media. Effective CFC destruction by steam plasma generated in an atmospheric pressure DC discharge has also been demonstrated by Watanabe, Taira, and Takeuchi (2005). Similar to freons, plasmas can be effectively applied for reducing perfluorinated compounds (PFCs, e.g., CF_4 and SF_6) emissions from chamber cleaning and dielectric etching tools in the semiconductor industry (Nantel-Valiquette et al., 2006). Although the plasma PFC abatement is mostly organized today at relatively low pressures, increasing attention is being given to the use of atmospheric pressure discharges for this application.

4.9 Cleaning of SO_2 with Nonthermal Plasma

SO_2 is generally not classified as a VOC, but SO_2 emission causes acid rain and results in serious environmental disasters. SO_2 emissions are also a precursor to particulates in the atmosphere. Both are causes for concern over the environmental impact of the emission of SO_2. The major sources of SO_2 emission are coal-burning power plants, steelworks, nonferrous metallurgical plants, as well as oil refineries and natural gas purification plants. The SO_2 emission in air is usually high volume and low concentration; the SO_2 fraction is usually on the level of hundreds of parts per million, and the total flow rate of polluted air in one system can reach millions of cubic meters per hour. Oxidation of SO_2 in air, usually in the presence of a catalyst such as NO_2, to SO_3 results in the rapid formation of sulfuric acid. However, the kinetics of this process is limited by the low concentration of both SO_2 and NO_2 in air. Therefore, the formation of SO_3 and sulfuric acid from the exhausted SO_2 occurs not immediately in the stacks of the industrial or power plants but later in the atmosphere, ultimately resulting in acid rain. Nonthermal plasma can be used to stimulate oxidation of SO_2 into SO_3 inside the stack. This permits the collection of the sulfur oxides in the form of sulfates, for example, in the form of a fertilizer, $(NH_4)_2SO_4$, if ammonia is mixed to the plasma-assisted oxidation products.

One of the commonly used approaches is the plasma-stimulated liquid-phase chain oxidation of SO_2 in droplets. Atomic oxygen and other reactive oxygen species generated in nonthermal air plasma are able to oxidize SO_2 to SO_3. The energy cost for the production of these species is about 10 eV per particle. Meanwhile, the SO_2 oxidation energy costs about 1–3 eV per molecule, which can be explained by the plasma stimulation of chain oxidation processes. There is no chain mechanism of SO_2 oxidation to SO_3 in conventional gas-phase chemistry without a catalyst. However, plasma is able to stimulate the chain oxidation of SO_2 either in liquid or humid air. Plasma-stimulated liquid-phase chain oxidation of SO_2 into SO_3 occurs in droplets formed in nonthermal discharges by water condensation around H_2SO_4, ions, and other active species. When the nonthermal plasma is generated in atmospheric air containing sulfur compounds, the active species (especially sulfuric acid) generated in the system immediately lead to water condensation and formation of mist. The presence of water permits the organization of a further chain process of SO_2 oxidation to SO_3 and sulfuric acid in the liquid phase. Liquid-phase chain oxidation of SO_2 into SO_3 and sulfuric acid has been investigated by several groups, including Calvet (1985), Penkett and Jones (1979), Daniel and Jacob (1986), and Deminsky et al. (1990). The oxidation mechanism in droplets depends on their acidity, which can be presented through the kinetic pathways discussed next.

4.9.1 Acidic Water Case (pH < 6.5)

The major ion formed in a water solution by gaseous SO_2 is the HSO_3^- ion:

$$SO_2 + H_2O \rightarrow HSO_3^- + H^+ \qquad (4.26)$$

The product of oxidation in this case is the sulfuric acid ion, HSO_4^-. The oxidation of HSO_4^- in the acidic solution into HSO_3^- is a chain process starting

TABLE 4.4

Chain Reaction of Oxidation Process
of SO_2 in Acidic Water Solution

Reaction	k (cm³/s)
$SO_3^- + O_2 \rightarrow SO_5^-$	2.5×10^{-12}
$SO_5^- + HSO_3^- \rightarrow SO_4^- + HSO_4^-$	1.7×10^{-16}
$SO_5^- + SO_5^- \rightarrow 2SO_4^- + O_2$	1.0×10^{-12}
$SO_4^- + HSO_3^- \rightarrow HSO_4^- + SO_3^-$	3.3×10^{-12}
$SO_5^- + HSO_3^- \rightarrow HSO_5^- + SO_3^-$	$4.2 \text{ v } 10^{-17}$
$HSO_5^- + HSO_3^- \rightarrow 2SO_4^- + 2H^+$	2.0×10^{-14}

Source: Fridman, A. (2008) *Plasma chemistry*, Cambridge University Press, Cambridge.

from the chain initiation through the HSO_3^- decomposition in collision with an active particle M (in particular, an OH radical):

$$M + HSO_3^- \rightarrow MH + SO_3^- \quad (4.27)$$

Chain propagation in the acidic solution starts with the attachment of oxygen and proceeds through active sulfur ion radicals SO_3^-, SO_4^-, and SO_5^-. The involved reactions are summarized in Table 4.4.

Termination of chain SO_2 oxidation is due to the recombination and destruction of the active sulfur ion radicals SO_3^-, SO_4^-, and SO_5^-:

$$SO_3^- + SO_3^- \rightarrow S_2O_6^{2-} \quad (4.28)$$

$$SO_5^- + SO_5^- \rightarrow S_2O_8^{2-} + O_2 \quad (4.29)$$

$$SO_3^-, SO_4^-, SO_5^- + \text{admixture} \rightarrow \text{destruction}$$

4.9.2 Neutral and Basic Water Cases (pH > 6.5)

The major ion formed in a water solution by gaseous SO_2 is the SO_3^{2-} ion:

$$SO_5^- + SO_5^- \rightarrow S_2O_8^{2-} + O_2 \quad (4.30)$$

The product of SO_2 oxidation in these cases is the sulfuric acid ion, SO_4^{2-}. The oxidation of SO_3^{2-} in neutral and basic solutions into the sulfuric acid ion

TABLE 4.5

Chain Reaction of Oxidation Process of SO_2 in Neutral and Basic Water Solutions

Reaction	k (cm³/s)
$SO_3^- + O_2 \rightarrow SO_5^-$	2.5×10^{-12}
$SO_5^- + SO_3^{2-} \rightarrow SO_4^{2-} + SO_4^-$	5.0×10^{-14}
$SO_4^- + SO_3^{2-} \rightarrow SO_4^{2-} + SO_3^-$	3.3×10^{-12}
$SO_5^- + SO_3^{2-} \rightarrow SO_5^{2-} + SO_3^-$	1.7×10^{-14}
$SO_5^{2-} + SO_3^{2-} \rightarrow 2SO_4^-$	2.0×10^{-14}

Source: Fridman, A. (2008) *Plasma chemistry*, Cambridge University Press, Cambridge.

is also a chain process. It starts from the chain initiation through the SO_3^{2-} decomposition in collision with an active particle M, in particular an OH radical, and the formation of active SO_3^{2-} radical as given by the following reaction:

$$M + SO_3^{2-} \rightarrow M^- + SO_3^- \tag{4.31}$$

Chain propagation in the neutral and basic solutions also starts with the attachment of oxygen and after that proceeds through the active sulfur ion radicals SO_3^-, SO_4^-, and SO_5^- as summarized in Table 4.5.

The chain termination reactions in the cases of neutral and basic solutions are mostly the same as in the case of acidic solution. The SO_3^-, ion radical dominates the chain propagation in both low- and high-acidity cases. When oxygen is in excess in the solution, the SO_3^-, ion radicals can be significantly replaced by SO_5^- ion radicals in the propagation of the oxidation chain reaction.

5

Biological Applications

5.1 Plasma Water Sterilization

Pulsed plasma technology is a cost-effective and environmentally friendly technology for the destruction of microorganisms in contaminated potable water and wastewater (Moisan et al., 2002; Akishev et al., 2006; Locke, Sato, et al., 2006; Anpilov et al., 2002; Sun et al., 1998; Sunka et al., 1999). Pulsed plasma discharges in water can produce high electric field, strong ultraviolet (UV) radiation, shock waves, ozone, H_2O_2, short-living active species (OH, H, O, $^1O_2$, HO_2, O_2^-), and charged particles for the effective sterilization of water (Fridman, 2008). There are two main approaches for using high-voltage pulse discharges for water sterilization; one is to use high-energy pulses that are 1 kJ or higher (Ching et al., 2001), and the other is to use low-energy pulses of about 1 J (Anpilov et al., 2002; Anpilov et al., 2004). The second approach is especially interesting due to the simplicity in designing the discharge system, which can readily be integrated into a household water delivery system. The second approach can also be developed into a portable water supply system for safe drinking water in remote areas or underdeveloped countries. Next, advances in plasma applications for water sterilization are reviewed.

5.1.1 Previous Studies of Plasma Water Sterilization

Schoenbach and his colleagues at Old Dominion University (Norfolk, VA) studied the feasibility of the use of electrical pulses in a microsecond range for the sterilization of biological cells for the past two decades (Abou-Ghazala et al., 2002; Schoenbach et al., 1997; Joshi, Hu, et al., 2002; Joshi, Qian, and Schoenbach, 2002; Zhang et al., 2006). For example, a 600-ns, 120-kV square wave pulse was used to generate pulsed corona discharge for bacterial (*Escherichia coli* and *Bacillus subtilis*) decontamination of water. They reported three orders of magnitude reduction in the concentration of *E. coli* (gram-negative bacterium) with an energy expenditure of 10 kJ/L. For *B. subtilis* (gram-positive bacterium), it took about 40 kJ/L to obtain the three-order reduction. However, it was observed that plasma had no effect on *B. subtilis* spores, a phenomenon attributed to the accumulation of electrical charges at the membrane of the spores, shielding the interior of the cell from the external electrical fields. Since the typical time required to charge mammalian cell membranes is about 1 μs (Schoenbach et al., 2004), microsecond pulses could not penetrate into cells once the membrane was charged. Hence, a short pulse in a nanosecond range might be able to penetrate the entire cell, nucleus, and organelles and affect cell functions, thus disinfecting them, a hypothesis that needs to be validated.

Researchers at the Eindhoven University of Technology in the Netherlands applied pulsed electric fields and pulsed corona discharges to inactivate microorganisms in water (Heesch et al., 2000). They utilized four different types of discharge configurations: (1) a perpendicular water flow over two wire electrodes; (2) a parallel water flow along two electrodes; (3) air bubbling through a hollow-needle electrode toward a ring electrode; and (4) a wire cylinder with the application of 100-kV pulses with a 10-ns rise time and 150-ns pulse duration. Inactivation rate was 85 kJ/L per one-log reduction for *Pseudomonas fluorescens* (gram-negative bacterium) and 500 kJ/L per one-log reduction for *Bacillis sereus* spores.

Researchers at the General Physics Institute, Russian Academy of Science, Moscow, studied if plasma systems could eradicate microorganisms such as *E. coli* and coliphages in water distribution systems (Anpilov et al., 2001, 2002). They reported that the active species, UV, ozone, and hydrogen peroxide effectively could sterilize bacteria in water.

Sato and his colleagues at Gunma University, Japan, investigated the feasibility of using plasma discharges for sterilization and removal of organic compounds in water (Sato et al., 1996; Sugiarto, Ohshima, and Sato, 2002; Sun, Sato, & Clements, 1999; Sugiarto et al., 2003). In particular, they studied the formation of chemical species and their effects on microorganisms and reported that hydroxyl radicals had an extremely short lifetime of 70 ns and diffused only 20 nm before they were absorbed in water. They also reported that hydrogen peroxide was produced through the recombination of hydroxyl radicals rather than by electrolytic reaction. They measured the emission spectrum between 200 and 750 nm and found that the largest peaks were in the UV range, which were believed to be molecular emissions from hydroxyl radicals.

Researchers at the University of Wisconsin, Madison, studied the feasibility of using dense-medium plasma reactors for the disinfection of water (Manolache et al., 2001). They found that the UV radiation emitted from the electrohydraulic discharge was the lethal agent for the inactivation of *E. coli* colonies rather than the thermal/pressure shocks or other active chemical species.

Akiyama and his colleagues at Kumamoto University, Japan, studied the possibility of using streamer discharges in water with a Marx generator to produce high-energy electrons, ozone, other chemically active species, UV radiation, and shock waves (Akiyama, 2000; Katsuki et al., 2002; Lisitsyn et al., 1999a). A thin-wire electrode placed coaxial with a cylindrical electrode was used to produce a large-volume discharge needed for industrial water treatments. Since the influence of the electric conductivity of water was found to be small, they speculated that bulk heating via ionic current did not contribute to the initiation of the breakdown process.

The previous research strongly suggested that plasma discharge has the ability to effectively inactivate microorganisms in water. A plasma-based water treatment system has a number of advantages compared to chemical or mechanical water treatment methods, such as minimal maintenance, low operating power, and minimal pressure loss through the plasma discharge device. Therefore, plasma-based water treatment can be implemented as a point-of-use water treatment system and in a large industrial water treatment system. Commercial systems to apply plasma discharges directly to water, typically known as high-energy pulsed plasma (HEPP) systems, are available from Dynawave and Actix. Compared to the nonthermal plasma discharges, these systems for water treatment have the following drawbacks:

1. Thermal plasma (high-temperature plasma): This poses a high risk of electrode erosion and is not very effective in the generation of radicals.

2. Very high pressure: It requires high-pressure protection, a safety concern when applied to an ordinary piping system.

3. Use of chemicals: These systems use small amounts of coagulant and polymer. Continuous supply of such chemicals can be a problem if the plasma water treatment is for a point-of-use application.

4. Large residence time in a clarifier: A large treatment tank is required, which is not practical for many applications.

5.1.2 New Developments in Plasma Water Sterilization

The present section examines the sterilization efficiency of three different plasma discharge prototypes (i.e., point-to-plane electrode configuration, magnetic gliding arc [MGA], and elongated spark discharge). Spark gap generator was used to produce a pulsed voltage capable of initiating the desired pulsed plasma discharge.

5.1.2.1 Point-to-Plane Electrode Configuration

The first configuration utilized a point-to-plane electrode geometry. Initial experiments included stainless steel and tungsten wire electrodes of varying diameters (0.18–2.5 mm). Variance in the plasma from corona to spark discharge was observed to be dependent on the gap distance between the anode and the grounded cathode.

Electrodes that were both rigid and electrically insulated were fabricated. This design included a stainless steel electrode (0.18 mm) encased in silicone residing in a hollow Teflon tube that was inserted in a glass tube, providing

the necessary insulation for the electrodes. The stainless steel wire was cho-
sen as the electrode due to its high melting temperature and relative stability
at high temperature. The stainless steel electrode extended approximately 1
mm beyond the tip of the glass tube, providing a region for spark discharge
initiation.

The critical gap distance where both spark discharge and corona discharge
exist was observed to be approximately 50 mm between the two electrodes.
When the gap distance was greater than 50 mm, a corona discharge was
obtained, whereas when it was less than 50 mm, a spark discharge was
obtained. When the distance between the two electrodes was small, the high
voltage initiated a channel breakdown in water, thus leading to a spark discharge. Conversely, when the distance between the two electrodes increased
and reached the critical distance, the plasma ceased growing, and a corona
discharge initiated.

Figure 5.1 presents voltage, current, and power profiles measured during a
typical pulsed spark test. The initial steep rise in the voltage profile indicates
the time moment of breakdown in the spark gap, after which the voltage
linearly decreased with time over the next 17 µs due to a long delay time
while the corona discharge was formed and transferred to a spark. The rate
of the voltage drop over time depended on the capacitance used. The duration of the initial peak was measured to be approximately 70 ns. At $t = 17$
µs, there was a sudden drop in the voltage, indicating the onset of a spark
or the moment of channel appearance, which was accompanied by sharp
changes in both the current and power profiles. The duration of the spark
was approximately 2 µs, which was much longer than the duration of the
corona.

The bacteria selected for the biological validation test was a nonpatho-
genic (i.e., noninfectious) strain of *E. coli*, which was considered the most
reliable measure of public risks in drinking water since its presence was an
indicator of fecal pollution and the possible presence of enteric pathogens.
Bacterial growth and measurement techniques included the production
of agar plates, incubating and growing bacteria, and performing bacterial
colony counts on the plates. This method of counting bacteria colonies is
a widely accepted practice in biology, called the *heterotrophic plate counting
method*. The complete procedure for the growth and utilization of *E. coli*
used in the present study can be found elsewhere (Madigan and Martinko,
2006). The results of the biological validation tests are given in Figures 5.2
and 5.3 for two different initial conditions. When the initial cell count was
relatively low (i.e., 10^6 CFU/mL), the spark discharge could produce a total
six-log reduction at an energy cost of 80 J/L per log. When the initial cell
count was high (i.e., 10^8 CFU/mL), the spark discharge produced a four-log
reduction at 800 J/L.

In an attempt to optimize the spark discharge system for effective water
sterilization, the energy per pulse was changed using different values of
capacitance, which was achieved by varying the number of capacitors in

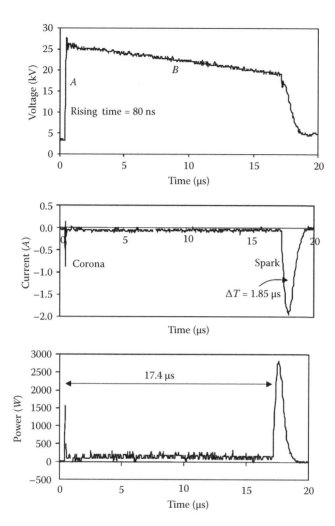

FIGURE 5.1

Voltage, current, and power profiles measured during a typical pulsed spark test.

the capacitor bank. Figure 5.4 shows the survival plot of *E. coli* for different energies per pulse. The D value obtained for an energy per pulse of 1.7 J was 187 J/L. Note that the D value is defined as the energy required to achieve one-$\log_{10}$ reduction in bacterial concentration at a specific plasma treatment condition. A low D value of 98 J/L was obtained for an energy per pulse of 1 J. It may be assumed that only a portion of the energy input into water contributes to the inactivation of microorganisms, while the rest of the energy is dissipated into the water in the form of heat. For energies per pulse of 0.68 J and 0.34 J, D values were 140 and 366 J/L, respectively. The optimized treatment system corresponded to a minimum D value of 187 J/L and energy per pulse of 1 J.

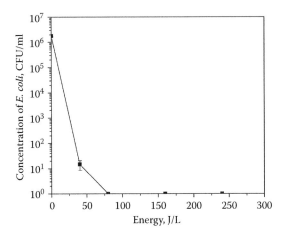

FIGURE 5.2
Survival plot obtained for an *E. coli* concentration of 10^6 CFU/mL.

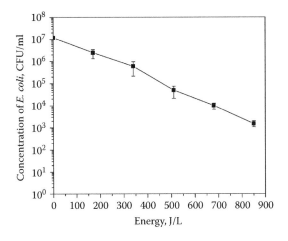

FIGURE 5.3
Survival plot obtained for an *E. coli* concentration of 10^8 CFU/mL.

5.1.2.2 Magnetic Gliding Arc Configuration

An MGA utilizes a constant direct current (DC) voltage to create an arc discharge between two coaxial electrodes. The arc discharge moves (or glides) with the help of an externally applied magnetic field. The schematic diagram given in Figure 5.5 shows the design and operational principle of the MGA. The MGA was made up of concentric electrodes with a central high-voltage cathode (1) and a grounded coaxial cylindrical anode (2). Multiple ceramic ring magnets were oriented such that one had an axial magnetic field within the grounded coaxial cylindrical anode as indicated by dotted arrows (5). A spiral wire (3) was attached to the cathode and was arranged close to the

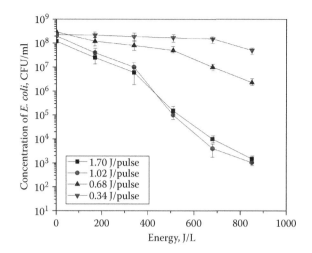

FIGURE 5.4
Survival plot of *E. coli* for different energies per pulse.

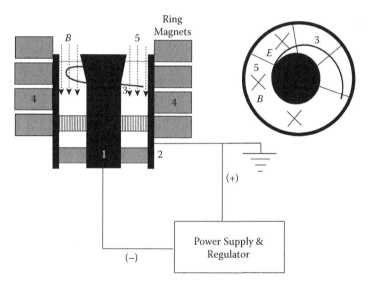

FIGURE 5.5
Sketch of a magnetic gliding arc. Multiple ring magnets were used to form a magnetic field B, which paralleled the axial direction of the reaction vessel. Electric field E was in the radial direction of the reactor.

anode to initiate the breakdown. The liquid to be treated was introduced from the top of the reactor and fell to the bottom exit by gravity.

The arc discharge after initiation was rotated by the magnetic field along the spiral wire and was forced to stabilize in a form that looked like a "plasma disk." The Lorentz force can be described by the cross product of the charge velocity and magnetic field, and the direction of the force can be determined using the right-hand rule:

$$\vec{F} = q\vec{V} \times \vec{B} \tag{5.1}$$

Therefore, the Lorentz force continued to rotate the arc around the cylindrical vessel.

After the initiation of the arc discharge using the spiral wire extended from the center electrode, there was an arc, at any given time, between the central cylindrical cathode and the outer ring anode. This MGA plasma could treat water quasiuniformly as the water moved along the spiral trajectory over the internal surface of the anode and fell by the gravity. The power supply for the MGA utilized DC rather than pulsed discharge. A reactive-resistance-capacitance power supply was developed such that the internal reactive resistance mimicked a serial active resistance, and the efficiency of power transfer to the plasma was close to 100%.

Figure 5.6 shows the photographs of the MGA in operation together with an introduction of water flow into the gliding arc reactor. A high-speed camera was used to capture the motion of the arc, which looked like a disk to the naked human eye. Photographs were taken with shutter speeds of 1/30 to 1/640 s. At a shutter speed of 1/640 s, one can clearly see one arc frozen between the two electrodes, whereas at a shutter speed of 1/30 s, one can see a plasma disk and multiple bright dots of the electrode spots in the places where they made a short interim stop.

A bacteria-laden solution was introduced into the vessel near the top of the anode cylinder and moved tangentially along the wall, creating a complete circle prior to entering the plasma region. After the solution was treated by

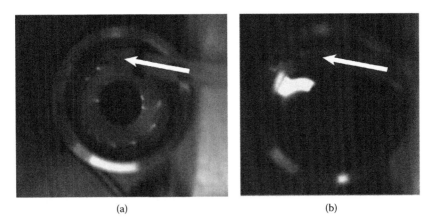

(a)   (b)

FIGURE 5.6
Magnetic gliding arc in operation (white arrows indicate water entry): (a) 1/30-s shutter speed; (b) 1/640-s shutter speed.

TABLE 5.1

pH and Temperature Variations in
Water after Plasma Treatment with
a Rotating Magnetic Gliding Arc

Plasma Power (W)	pH	T (°C)
0	7.23	23.8
121	6.65	24.6
182	6.64	25.2
239	6.67	25.5
299	6.36	25.4

TABLE 5.2

Biological Validation Experiment: Bacterial
Concentration after Magnetic Gliding Arc
Treatment

Power (W)	Cell Concentration (CFU/mL)		
	Run A	Run B	Run C
0	2.71E+06	3.60E+06	2.79E+06
121	0.00E+00	0.00E+00	0.00E+00
182	0.00E+00	0.00E+00	0.00E+00
239	0.00E+00	0.00E+00	0.00E+00
299	0.00E+00	0.00E+00	0.00E+00

the MGA, it descended to the bottom of the apparatus and was expelled through two exit ports on the side of the bottom anode cylinder.

It is important to note that the exposure of water to the gliding arc induced an initial decrease in pH. Table 5.1 shows the results of the measurements, where the pH value changed once the plasma was turned on. No significant fluctuation in water temperature was observed over variances in power levels. Table 5.2 shows the results of biological validation tests using the MGA; the number of *E. coli* was counted as a function of the applied power. At a power level of 121 W, the MGA could completely kill all *E. coli* in water, with a six-log reduction.

5.1.2.3 Elongated Spark Configuration

Long-spark ignition is a process of taking a single spark and elongating it through a series of capacitors. The long spark has potentially a significant advantage over the standard system with an increased spark gap. To increase the spark gap distance, the standard system should have both an increased capacitance and an increased voltage to initiate breakdown. The present long-spark technology only requires an individual capacitor per

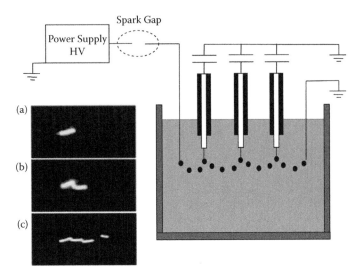

FIGURE 5.7
Schematic diagram of the long-spark discharge: (a) single ignition; (b) double ignition; (c) triple ignition. HV, high voltage.

adjacent electrodes, thus eliminating the need for increased supply voltage. The present system utilized a length of coaxial cable to create a series of high-voltage capacitors. The capacitance of the cable was 93.5 pF/m, with each cable length measuring approximately 0.9 m. The discharge was sustained over three of the capacitors with an overall spark length of approximately 25 mm (see Figure. 5.7).

5.1.3 Plasma Species and Factors for Sterilization

Plasma sterilization is a complicated process determined by multiple plasma species and factors, including thermal effect, charged particles, excited species, reactive neutrals, and UV radiation. The contribution of these factors differs in different types and regimes of plasma discharges. It also differs from the point of view of induced biological pathways, and the synergetic nature of the interactions between the plasma factors is important.

The first factor that usually comes to mind when discussing the sterilization of liquid is the thermal effect. Many conventional sterilization methods are based on the use of heat, with a typical treatment time of about 1 h. Note that most nonthermal plasma discharges operate at low temperatures and do not provide thermal sterilization. However, one should keep in mind that even strongly nonequilibrium discharges can be characterized by elevated temperatures in some localized intervals in time or in space. For example, dielectric barrier discharge (DBD) is traditionally considered nonthermal. But, the temperature inside DBD microdischarge channels can reach several hundred degrees, where the thermal effect should be taken into account. Overheating of the microdischarge channels occurs usually due to nonuniformity of electrodes, and the strong contribution of radicals or charged species to sterilization can also be enhanced by even a small temperature increase.

Next, consider the contribution of charged particles (electrons and ions) in plasma sterilization. The charged particles play a major role in sterilization, especially in the case of so-called direct plasma treatment, in which plasma is in direct contact with a microorganism. The effectiveness of electrons in water sterilization is related in particular to their behavior when they reach and become immersed in water, becoming hydrated electrons e_{aq}. The hydrated electrons are surrounded by polar water molecules and remain stable, providing the deep penetration of the electrons in water. The electrons react with different admixtures in water, in particular with biomolecules and peroxide (converting the latter into OH and OH⁻). Between the most important processes, a hydrated electron e_{aq} is able to convert oxygen dissolved in water into superoxide, the highly reactive O_2^- ion radical:

$$e_{aq} + O_2(H_2O) \rightarrow O_2^-(H_2O) \tag{5.2}$$

The superoxide is a precursor to other strong oxidizing agents, including singlet oxygen and peroxynitrite. It is interesting to mention that superoxides are produced in the human body naturally as an antibiotic agent to kill bacteria. In an aqueous system with a sufficient level of acidity, the superoxide is converted into hydrogen peroxide and oxygen in a reaction called the superoxide dismutation:

$$2O_2^- + 2H^+ \rightarrow H_2O_2 + O_2 \tag{5.3}$$

The superoxide dismutation can be spontaneous or can be catalyzed and therefore significantly accelerated by the enzyme superoxide dismutase. The superoxide not only generates hydrogen peroxide but also stimulates its conversion into OH radical, which is an extremely strong oxidizer and effective in sterilization.

The major negative ion in nonthermal plasma is O_2^-. When this ion is immersed in water, it becomes superoxide and further forms and OH according to Equation 5.3, similar to the case of sterilization provided by the chemical effect of plasma electrons. The chemical effect of positive plasma ions M^+ (N_2^+ e.g.,) can be illustrated by the process discussed next.

First, the interaction of the plasma ion M^+ with water molecules starts with fast charge transfer processes to water molecules, characterized by a relatively low ionization potential:

$$M^+ + H_2O \rightarrow M + H_2O^+ \tag{5.4}$$

Water ions then rapidly react with water molecules, producing and OH radicals:

$$H_2O^+ + H_2O \rightarrow H_3O^+ + OH \tag{5.5}$$

The positive plasma ions make water acidic. Thus, nonthermal plasma interaction with water conventionally decreases pH, a process that enhances the dismutation process of the superoxides through Equation 5.3. The contribution of both electrons and ions (positive and negative) in one way or another leads to the effective generation of OH and other highly reactive oxidants, resulting in the intensive oxidation of biomaterials and a strong sterilization effect.

Plasma is a source of UV radiation at multiple wavelengths, which can be effective in sterilization. UV radiation generally can be divided into four ranges: vacuum UV (VUV) radiation with a wavelength between 10 and 100 nm; UVC radiation between 100 and 280 nm; UVB radiation between 280 and 315 nm; and UVA radiation between 315 and 400 nm. VUV photons have high energy sufficient for breaking chemical bonds in biological cells. Their efficiency in sterilization, however, is limited due to a very short penetration depth. The efficiency of UVA and UVB photons in sterilization is limited by the low energy of the photons. UVC photons have energy sufficient for causing lethal damage to organic tissues and possess a large penetration depth at the same time, thus making UVC the most effective in direct sterilization processes.

Figure 5.8 shows measured optical emission spectra from an underwater corona discharge. The spectra indicated combinations of a continuum radiation and strong atomic line emissions due to the presence of hydrogen and oxygen radicals resulting from the decomposition of water. The most intense peak was the hydrogen Balmer series alpha line (H_α) at 656 nm. A broadened H_β line was also visible at 484 nm. In the infrared region, the four distinct lines could be observed due to atomic oxygen, with the strongest emission at 777 nm. In the UV range, only the OH line at 306 nm and the continuous decay of the H_β line were observed. Of note is that no significant lines due to metal ablation from the electrodes were present in the spectra. Significant

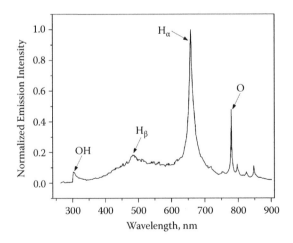

FIGURE 5.8
Typical optical emission spectra from corona discharge in water.

broadening of the H_α line was observed, which could mainly be attributed to Stark broadening, indicating a relatively high electron density inside the filament of the corona discharge (Bruggeman et al., 2009; Bruggeman et al., 2010).

To test the role of UV radiation in the inactivation of microorganisms, soluble sunscreen 2,2'-dihydroxy-4,4'-dimethoxybenzophenone-5,5'-disulfonic acid (BP-9) was added in water. BP-9 is essentially photostable because its excited reactive state undergoes extremely rapid internal conversion into the ground state, thereby degrading electronic into thermal energy before any photofragmentation occurs. The transmission spectrum for wavelengths between 200 and 300 nm was obtained for BP-9 solutions of concentrations ranging from 1 to 3,000 mg/L. Figure 5.9 shows the relative transmittance for various concentrations of the BP-9 solution. Intensities were normalized at 500 nm. It can be seen that the BP-9 solution of a concentration of 3 mg/L absorbed some portion of UVC but transmitted a major portion of UVB and UVA. The BP-9 solution of a concentration of 30 mg/L absorbed a significant amount of UVC, UVB, and some portion of UVA. The BP-9 solution of a concentration of 3,000 mg/L completely absorbed UVA, UVB, and UVC. The concentrations 3, 30, and 3,000 mg/L were hence selected for conducting the present inactivation experiments. As a control, the sensitivity of *E. coli* to BP-9 was tested by exposing the bacteria with an initial concentration of 1.44×10^7 CFU/mL to a BP-9 solution with a concentration of 7,500 mg/L of water for 2 h. No change in bacterial concentration was observed.

Figure 5.10 shows the survival curves for various concentrations of BP-9. It can be observed that the disinfection of *E. coli* in water was almost completely

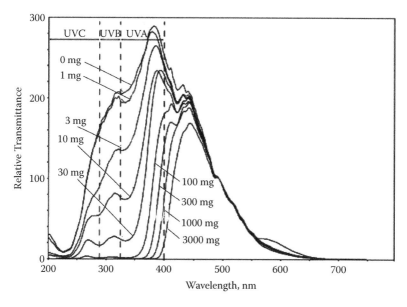

FIGURE 5.9
Fluorescence spectra for different concentrations of BP-9 solution.

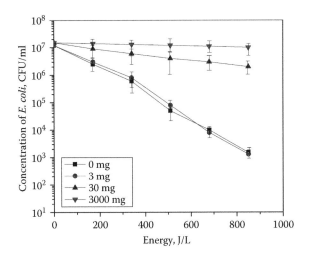

FIGURE 5.10
Survival plot of *E. coli* for various concentrations of BP-9.

suppressed at a concentration of 30 mg/L of BP-9. These results are in agreement with the results of similar experiments done earlier by other researchers (Ching, Colussi, and Sun, 2003). UV radiation is known to kill microorganisms in water (Hayamizu, Tenma, and Mizuno, 1989; Wolfe, 1990). DNA has an important absorption peak near 254 nm (Labas et al., 2006). UV photons can cause irreparable damage to the bacterial DNA, thus inactivating them (Moisan et al., 2001; Sun et al., 1999). The present results indicate that UV radiation produced by spark discharge in water plays a major role in inactivating microorganisms.

5.1.4 Comparison of Different Plasma Discharges for Water Sterilization

The possible applications for plasma water sterilization span a wide range of areas and industries, including foreign aid and disaster relief, providing a means for developing countries to sustain potable water sources, integrating into household water delivery systems in developed countries, and improving power plant wastewater treatment. Table 5.3 shows data comparing different plasma discharges found in plasma-based water treatment methods. The typical discharges applied in the water treatment can be found at two ends of the spectrum: either a highly energetic thermal discharge (i.e., arc discharge) or a less-energetic nonthermal discharge (i.e., corona discharge). The pulsed spark discharge used in the present study belongs somewhere between these two extremes. The properties of the spark discharge are unique and beneficial with regard to water treatment. First and most important, the pulsed spark discharge in the present experiments required a very low power in comparison to other systems. To achieve a four-log reduction for a typical household flow rate of 6 gpm (gallons per minute; 22.7 L/min), the electrical energy requirement

TABLE 5.3

Comparison of Different Plasma Discharges Used in Plasma-Based Water Treatment Methods

	Pulsed Arc Discharge	Pulsed Spark Discharge	Pulsed Corona Discharge
Energy per liter for 1-log reduction of *E. coli*	860 J/L	77 J/L	30–150 kJ/L
Power requirement for household (6 gpm)	0.326 kW	0.148 kW	11.4–56.8 kW
Power requirement for village (1,000 gpm)	54.3 kW	4.9 kW	1,892–9,463 kW
Efficiency	Good	Excellent	Poor
Maximum power available in small power system (10 × 10 × 10 cm overall size)	30 kW	10 kW	0.3 kW
Maximum water throughput based on maximum power	553 gpm	2058 gpm	0.03–0.16 gpm
Central lethal biological agent of discharge	UV and chemical radicals	UV and chemical radicals	Chemical radicals

Note: 1 gpm (gallon per min) = 3.786 liters per min.

was only 120 W. The low energy consumption enables the plasma water treatment system to be powered by solar cells or hand-crank power generators in third-world countries, where electricity is a scarce commodity. Second, the pH and temperature of the bulk water did not significantly change during the treatment, indicating that the energy of the discharge was not wasted in the form of bulk heating of water and was used effectively for the treatment. Third, the flow rates that can be treated by the present plasma discharge (i.e., pulsed sparks) could be significantly larger than with other plasma systems.

The present pulsed spark discharge system indicates a potential to accommodate a 1,000-gpm water flow rate while retaining the capability of achieving a four-log reduction in biological contaminant at a power of only 20 kW. This can be an extraordinary breakthrough in plasma water treatment. Plasma technology can be useful in the area of water treatment and can perpetuate new and improved methods of delivering potable water both nationally and internationally.

5.2 Blood Treatment Using Nonthermal Plasma

Blood coagulation is an important issue in medicine, particularly regarding wound treatment. Thermal plasma has been traditionally used for this application in the form of the so-called cauterization devices: argon plasma coagulators (APCs), argon beam coagulators, and so forth. In these devices, widely used in surgery, plasma jet is a source of local high-temperature heating, which cauterizes the blood (Vankov and Palanker, 2007). Recent developments of the effective nonthermal plasma medical systems permit effective blood coagulation to be achieved through nonthermal plasma stimulation of specific natural mechanisms in blood without heating and any damage to surrounding tissues (Laroussi, 2008; Kong et al., 2009). Both coagulating the blood and preventing the coagulation could be needed, depending on the specific application. In wound treatment, one would want to close the wound and sterilize the surrounding tissue. Flowing blood in that case would prevent wound closure and create the possibility of reintroduction of bacteria into the wound. Thus, clearly, an understanding of the mechanisms of blood coagulation by nonthermal plasma is needed.

5.2.1 In Vitro Blood Coagulation Using Nonthermal Atmospheric Pressure Plasma

Dielectric barrier plasma was experimentally confirmed to significantly accelerate blood coagulation in vitro. Visually, a drop of blood drawn from a healthy donor and left on a stainless steel surface coagulates on its own in about 15 min, while a similar drop treated for 15 s by DBD plasma coagulates in under 1 min (see Figure 5.11). Increase of the viscosity of the blood samples in accordance with different treatment times was observed at shear rates ranging from 1 to 1,000 s^{-1} (see Figure 5.12).

DBD treatment of blood plasma leads to similar results (see Figure 5.13). In addition, a significant change in blood plasma protein concentrations was observed after plasma treatment of blood plasma samples from healthy

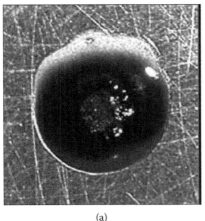

(a)   (b)

FIGURE 5.11
Blood drop treated by DBD: 15 s of DBD (left) and control (right). Photographs were taken 1 min after the drops were placed on brushed stainless steel substrate.

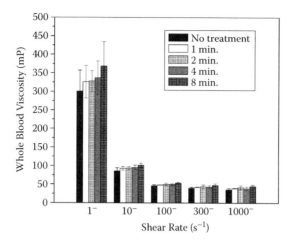

FIGURE 5.12
The increase of whole-blood viscosity according to different DBD treatment times.

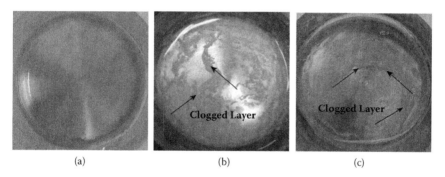

(a)   (b)   (c)

FIGURE 5.13
(See color insert.) The clotted formation of white layer in blood plasma sample with DBD treatment: (left) control; (middle) blood plasma treated with DBD for 4 min; (right) blood plasma treated with DBD for 8 min.

patients, patients with hemophilia, and blood samples with various anticoagulants. Anticoagulants, like sodium heparin, EDTA (ethylenediamine tetraacetic acid), and sodium citrate, are designed to bind various ions or molecules in the coagulation cascade, thus controlling the coagulation rate or preventing it altogether. Analysis of changes in the concentration of blood proteins and clotting factors indicated that DBD aids in promoting the advancement of blood coagulation; in other words, plasma is able to catalyze the biochemical processes taking place during blood coagulation.

5.2.2 In Vivo Blood Coagulation Using DBD Plasma

Plasma stimulation of in vivo blood coagulation was demonstrated by Fridman, Gutsol, and Cho (2007) in experiments with live hairless SKH1

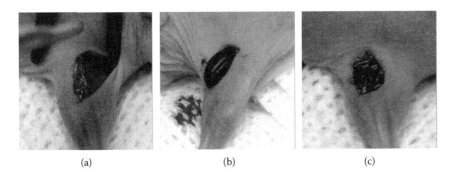

(a)   (b)   (c)

FIGURE 5.14
Blood coagulation of a live animal: (left) Saphenous vein is a major blood vessel for a mouse; (middle) if left untreated following a cut, animal will bleed out; (right) 15 s at 0.8 W/cm² stops the bleeding completely right after treatment.

mice. DBD plasma treatment for 15 s could coagulate blood at the surface of a cut saphenous vein and vein of a mouse (Figure 5.14). In these experiments, only the ability of direct nonthermal plasma treatment to coagulate blood was tested. A full in vivo investigation of the ability of plasma to accelerate wound healing through sterilization and blood coagulation was discussed by Fridman, Gutsol, and Cho (2007).

5.2.3 Mechanisms of Blood Coagulation Using Nonthermal Plasma

Detailed biochemical pathways of nonthermal plasma-stimulated blood coagulation remain largely unclear. Several possible mechanisms, however, were investigated (Fridman et al., 2006; Kalghatgi et al., 2007). It was demonstrated that direct nonthermal plasma could trigger natural, rather than thermally induced, coagulation processes. When the surface of blood was protected by thin aluminum foil, which prevented contact between blood and DBD plasma but transferred all the heat generated by the plasma, no influence on blood coagulation was observed, which proved the nonthermal mechanism of plasma-stimulated blood coagulation.

Natural blood coagulation is a complex process that has been studied extensively, and various nonthermal plasma products can affect this process at many of the coagulation steps. The initial plasma coagulation hypothesis was focused on increasing the concentration of Ca^{2+}, an important factor in the coagulation cascade of blood. Calcium ions circulate in blood in several different forms: 45–50% of calcium ions are free ionized, 40–45% are bound to proteins (mostly albumin), and the remaining calcium ions are bound to anions such as bicarbonate, citrate, phosphate, and lactate. Bound calcium ions are in dynamic equilibrium with free calciums. It was hypothesized that DBD was effective in the increase of Ca^{2+} concentration through a reduction/oxidation mechanism:

$$\left[Ca^{2+}R^{2-}\right]+H^{+}_{(aq)} \leftrightarrow \left[H^{+}R^{2-}\right]+Ca^{2+}_{(aq)} \tag{5.6}$$

provided by hydrogen ions generated through a sequence of ion molecular processes induced by plasma ions. (Here, R represents the calcium-binding protein complexes, such as S100A7 and albumin.) The validity of this hypothesis was tested by measuring the Ca^{2+} concentration in the DBD-treated anticoagulated whole blood using a calcium-selective electrode (Kalghatgi et al., 2007). Calcium concentration was measured immediately after DBD treatment for 5 to 60 s. No significant change occurred in calcium ion concentration during the typical time scale of blood coagulation in discharge-treated blood. In vivo, the pH of blood was maintained in a very narrow range of 7.35–7.45 by various physiological processes. The change in

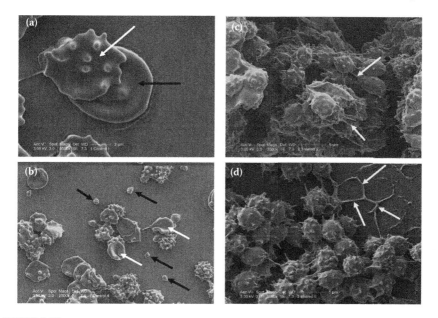

FIGURE 5.15
(a) Citrated whole blood (control) showing a single activated platelet (white arrow) on a red blood cell (black arrow); (b) citrated whole blood (control) showing many nonactivated platelets (black arrows) and intact red blood cells (white arrows); (c) citrated whole blood (treated) showing extensive platelet activation (pseudopodia formation) and platelet aggregation (white arrows); (d) citrated whole blood (treated) showing platelet aggregation and fibrin formation (white arrows). (From Kalghatgi, S., Fridman, G., Cooper, M., Nagaraj, G., Peddinghaus, M., Balasubramanian, M., Vasilets, V., Gutsol, A., Fridman, A., and Friedman, G. (2007) Mechanism of blood coagulation by nonthermal atmospheric pressure dielectric barrier discharge plasma. *IEEE Trans. Plasma Sci.* 35, 1559–1566.)

pH by plasma treatment was less than the natural variation of pH, indicating that the coagulation was not due to the Ca^{2+} concentration change induced by pH change in blood.

Instead, the evidence pointed to the selective action of direct nonthermal plasma on blood proteins involved in natural coagulation processes. DBD treatment of whole-blood samples was believed to change concentrations of proteins participating in the coagulation cascade. The final step in the natural biological process of blood coagulation was the production of thrombin, which converted fibrinogen into fibrin monomers that polymerized to form fibrin microfilaments. DBD plasma treatment of fibrinogen solution in physiological medium coagulated it, a process confirmed visually through a change in the color of the solution and through dynamic light-scattering measurement. Of note is that plasma did not influence fibrinogen through a pH or temperature change. To test the specificity of the DBD treatment on blood proteins, albumin, an important blood protein that did not participate in the coagulation cascade but is used as biological glue in some cases, was tested in the same fashion as the fibrinogen solution. No change was observed either visually or through dynamic light scattering, indicating that plasma was unable to polymerize albumin. Thus, it can be concluded that nonthermal plasma selectively affected proteins (specifically fibrinogen) participating in the natural coagulation mechanism.

Morphological examination of a clot layer of whole blood by scanning electron microscopy (SEM) further proved that plasma did not "cook" blood but initiated and enhanced natural sequences of blood coagulation processes. Activation of platelets followed by aggregation was the initial step in the coagulation cascade, and the conversion of fibrinogen into fibrin was the final step in the coagulation cascade. Figure 5.15 shows extensive platelet activation, platelet aggregation, and fibrin formation following DBD plasma treatment.

6

Cooling Water Treatment Using Plasma

6.1 Introduction

Water is used as a cooling medium in various industrial facilities, such as large centralized air-conditioning systems, and in thermoelectric power plants. In both cases, the cooling water plays an essential role in removing heat from condensers. According to the U.S. Geological Survey's (USGS) water use survey data (Feeley et al., 2008), thermoelectric generation accounted for 39% (136 billion gallons per day [BGD]) of all freshwater withdrawals in the nation in 2000, second only to irrigation (see Figure 6.1) (Feeley et al., 2008). Furthermore, the average daily national freshwater consumption for thermoelectric power generation is predicted to increase from the current 4 BGD for the production of approximately 720 GW electricity to 8 BGD for 840 GW in 2030 (see Figure 6.2) (Feeley, 2008).

In cooling water management, it is important to distinguish between water withdrawal and water consumption. Water withdrawal represents the total water taken from a source, while water consumption represents the amount of water withdrawal that is not returned to the source. Freshwater consumption for the year 1995 (the most recent year for which these data are available) is presented in Figure 6.3. Freshwater consumption for thermoelectric uses appears low (only 3%) when compared with other use categories (irrigation was responsible for 81% of water consumed). However, even at 3% consumption, thermoelectric power plants consumed more than 4 BGD (Feeley et al., 2008).

A modern 1,000-MW thermoelectric power plant with 40% efficiency would reject 1,500 MW of heat at the full load. This is roughly equivalent to 512×10^6 Btu/hr and uses about 760,000 gal/min of circulating water based on 10°C temperature difference in condenser (El-Wakil, 1984). As heat is removed via the evaporation of pure water at the cooling tower, the need for the makeup water is about 7,500 gal/min for a typical fossil plant, which results in 10 million gallons a day (El-Wakil, 1984).

One of the critical issues in cooling water management is condenser tube fouling by mineral ions such as calcium and magnesium. Since calcium carbonate $CaCO_3$ is the most common issue in cooling water, one can use the words *calcium scale* to refer to all scales caused by mineral ions. To prevent or minimize condenser tube fouling, the COC (cycle of concentration) in

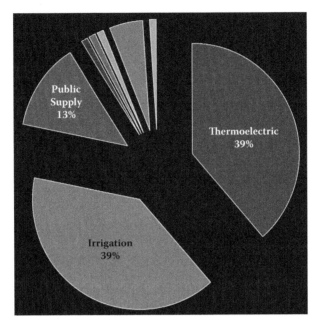

FIGURE 6.1
U.S. freshwater withdrawal in 2000. (Total may not be equal to 100% due to rounding.) (From U.S. Geological Survey, *Estimated use of water in the United States in 2000*, USGS Circular 1268, March 2004, U.S. Geological Survey, Denver.)

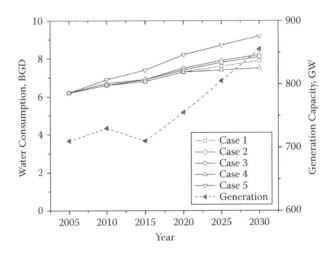

FIGURE 6.2
Average daily national freshwater consumption for thermoelectric power generation 2005–2030 (predicted). (From Department of Energy/Office of Fossil Energy's Energy and Water R&D Program, 2008.)

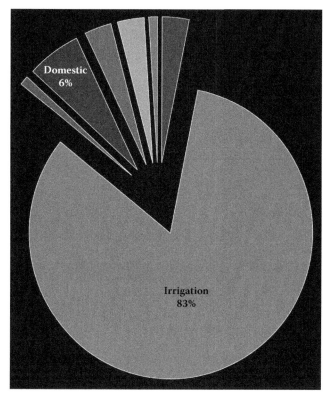

FIGURE 6.3

U.S. freshwater consumption in 1995. (Total may not equal 100% due to rounding.) (From U.S. Geological Survey, *Estimated use of water in the United States in 1995,* USGS Circular 1220, 1998 U.S. Geological Survey, Denver.)

wet recirculation cooling systems is often kept at 3.5. Since increasing the COCs can reduce the amount of makeup water, the water consumption can be reduced with the increased COCs. For example, if one can increase the COCs to 8, the freshwater consumption can be reduced by approximately 25%, meaning that the makeup water can be reduced by 2.5 million gallons a day in a 1,000-MW thermoelectric power plant.

Since the amounts of mineral ions in circulating cooling water primarily depend on the COC, condenser tube fouling also depends on the COC. Hence, the issue in cooling water management is to increase the COC without the condenser fouling problem. The present chapter deals with an innovative water treatment technology utilizing plasma discharges in water, with which one can increase the COC without the fouling problem in condenser tubes. The key issue is how to precipitate and remove mineral ions such as calcium and magnesium from circulating cooling water so that the calcium carbonate scales can be prevented at the condenser tubes and at the same time the COC can be increased.

In the next few sections, new developments of underwater plasma treatment at Drexel Plasma Institute for various applications are discussed.

6.2 Self-Cleaning Filtration Technology with Spark Discharge

In modern wastewater treatment, filters are routinely used for removing unwanted particles from water. Conventionally, microfiltration methods are used to remove suspended particles from water. Whenever a filter is used in a water system, the pressure drop across the filter gradually increases with time or the flow rate gradually decreases with time. This reduced performance of a filter is due to the accumulation of impurities on the filter surface, and the clogged area becomes a site for bacteria growth, further reducing the opening in the filter surface, increasing the pumping cost. Therefore, to remove suspended particles from water continuously, the filter must be replaced frequently, a process that is prohibitively expensive in most industrial water applications. To overcome the drawbacks of frequent filter replacement, self-cleaning filters are commonly used in industry. Although there are a number of self-cleaning filter technologies available on the market, most self-cleaning filters use a complicated backwash method, which reverses the direction of flow during the cleaning phase. Furthermore, the water used in the backwash must be clean, filtered water, which reduces the filter capacity. The aforementioned drawbacks of the conventional filter technologies motivated us to develop a new self-cleaning filter using spark-generated shock waves.

As illustrated in previous chapters, strong shock waves can be formed during the process of pulsed arc or spark discharge. The energy transferred to the acoustic energy can be calculated as (Lu, Pan, and Liu, 2002)

$$E_{acoustic} = \frac{4\pi r^2}{\rho_0 C_0} \int \left(P(r,t) - P_0 \right) dt \tag{6.1}$$

where r is the distance from the spark source to the pressure transducer, r_0 is the density of water, C_0 is the speed of sound in water, and P_0 is the ambient pressure. One can conclude that the pressure created by the spark discharge is much higher than the ambient pressure at positions close to the source. Traditionally, the high-pressure shock wave is studied for high-voltage insulation and rock fragmentation, while recently it has found more applications in other areas, including extracorporeal lithotripsy and metal recovery from slag waste (Bluhm, Frey, and Giese, 2000; Akiyama, Sakugawa, and Namihira, 2007; Wilson, Balmer, and Given, 2006; Snizhko et al., 2007; Sridharan et al., 2007).

To validate the concept of using spark discharge for filter cleaning, an experimental setup was built in which discharges could be produced in water, and the pressure drop across a filter surface was measured over time at various spark frequencies and flow conditions. It was hypothesized that the energy deposited by the spark shock wave onto the water-filter interface was enough to remove the contaminants having Van der Waals bonds with the filter surface. The objective of the study was to examine the feasibility of a self-cleaning water filtration concept using spark discharges in water.

Figure 6.4 shows the long-time response of the pressure drop across the filter surface after one single spark discharge at different flow rates. One could visually observe that some particles were dislodged from the filter surface and were pushed away from the filter surface by tangential flow, and a sudden change in the pressure drop immediately following the single spark discharge confirmed the removal of the deposits from the filter surface.

The cleaning effect can be explained by the pressure pulse produced by spark discharge. A number of researchers studied the bubble growth by spark discharge in water (Joseph and Miksis, 1980; Inoue and Kobayashi, 1993; Prosperetti and Lezzi, 1986; Lezzi and Prosperetti, 1987). One of the most effective models is the Kirkwood-Bethe model (Lu, Pan, and Liu, 2002):

$$\left(1 - \frac{\dot{R}}{C}\right) R \ddot{R} + \frac{3}{2}\left(1 - \frac{\dot{R}}{3C}\right) \dot{R}^2 = \left(1 + \frac{\dot{R}}{C}\right) H + \left(1 - \frac{\dot{R}}{C}\right) \frac{R}{C} \dot{H} \qquad (6.2)$$

where C and H are the speed of sound of the water and the specific enthalpy at the bubble wall, respectively. R is the radius of the bubble wall. The

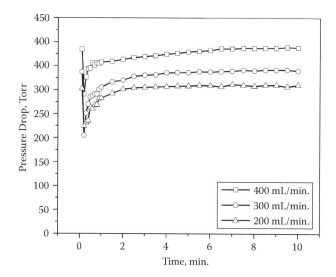

FIGURE 6.4
Variations of pressure drop after one single spark discharge at three different flow rates with an artificially hardened water. (From Yang, Kim, et al., 2009b.)

overdots denote the derivatives with respect to time. By expressing the time derivative of specific enthalpy as a function of the derivative of plasma pressure P inside the bubble, Lu and coworkers showed that it was possible to solve P as (Lu, Pan, and Liu, 2002)

$$P(r,t_r) = A\left[\frac{2}{n+1} + \frac{n-1}{n+1}\left(1 + \frac{n+1}{rC^2}G\right)^{1/2}\right]^{(2n/(n-1))} - B \qquad (6.3)$$

where A, B, and n are constants (A = 305.0 MPa, B = 304.9 MPa, n = 7.15), r is the distance from the source of the spark to the pressure transducer, and

$$G = R(H + \dot{R}\dot{R}/2), \ t_r = t + (r-R)/C_0 \qquad (6.4)$$

Using Equation 6.4, Lu et al. simulated that for a spark discharge with energy of 4.1 J/pulse, the maximum pressure at a distance of 0.3 m could be up to 7 atm (Lu, Pan, and Liu, 2002).

With a single pulse, it took approximately 3 min for the pressure drop to return to its asymptotic value after the application of the single spark discharge. This suggests that one needs to apply spark discharges repeatedly to remove the particles effectively from the filter surface over an extended period.

Figure 6.5 shows the changes in the pressure drop over time for three different flow rates. One spark discharge was applied every minute from the supply water side (i.e., untreated water side) where the accumulation of

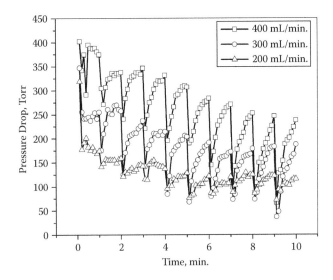

FIGURE 6.5
Changes in pressure drop under repeated pulsed spark discharges with an artificially hardened water. (From Yang, Kim, et al., 2009b.)

suspended particles took place. For the case of 300 mL/min, the pressure drop decreased from the maximum asymptotic value of 350 torr to 230 torr after the first spark discharge. Since water with particles was continuously circulated through the filter surface, the pressure drop began to increase immediately after the completion of the first spark discharge, as shown in Figure 6.4. The second and third spark discharges further reduced the pressure drop to 170 and 125 torr, respectively. The pressure drop again began to increase immediately after each spark discharge. The sixth spark discharge brought the pressure drop down to a value of approximately 65 torr, and subsequent spark discharges almost resulted in the minimum value of the pressure drop. For the cases of 200 and 400 mL/min, similar trends of the changes in the pressure drop were observed.

Figures 6.6a and 6.6b show the changes in the pressure drop under repeated pulsed spark discharges with frequencies of 2 and 4 pulses/min, respectively. Three horizontal arrows indicate the original asymptotic values for three different flow rates, which were the maximum pressure drop due to a clogged filter surface by calcium carbonate deposits. The first spark discharge significantly reduced the pressure drop in both cases. After that, the rate of the reduction slowed. The pressure drop oscillations observed after the application of spark pulses reached quasisteady conditions after about 10 pulses for both cases. In these oscillations, the maximum pressure drop decreased to about 45% of its original asymptotic value, while the minimum pressure drop was close to that of the clean filter. These results demonstrate

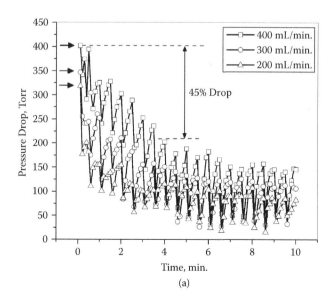

(a)

FIGURE 6.6
Changes in pressure drop under repeated pulsed spark discharges with frequencies of (a) 2 pulses/min and (b) 4 pulses/min. (From Yang, Kim, et al., 2009b.)

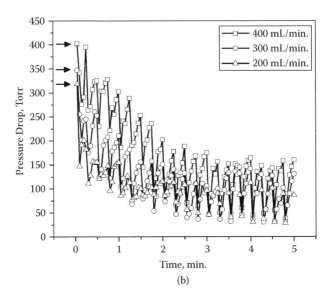

FIGURE 6.6 (CONTINUED)
Changes in pressure drop under repeated pulsed spark discharges with frequencies of (a) 2 pulses/min and (b) 4 pulses/min. (From Yang, Kim, et al., 2009b.)

the validity of the present spark discharge method. Note that the present cleaning method using the spark discharge does not require a backwash to remove deposits from the filter surface or stopping the flow. Furthermore, the present spark discharge method can maintain the pressure drop across the filter at a rather low value (i.e., almost close to the initial clean state), thus providing a means to save not only freshwater but also electrical energy both for the operation of pump and for the backwash in the conventional backwash system.

Figure 6.7 shows the changes in the pressure drop over time with the anode electrode placed beneath the filter membrane (i.e., plasma discharge was applied from the treated water side). In this case, the momentum transfer from the shock wave to particles on the filter surface was indirect and had to go through the membrane. Figure 6.7 clearly shows that the pressure drop did not improve significantly in this case, indicating that the cleaning effect was negligible compared with when the electrode was placed at the untreated water side. The fact that the momentum transfer from the shock wave to the membrane was weak is actually good news. The low energy transfer rate means that the present spark discharge may not deform the membrane significantly and therefore will not damage the membrane and has the potential to be applied in the cleaning of more delicate membranes, such as in a reverse osmosis as well as porous filters over an extended period of time.

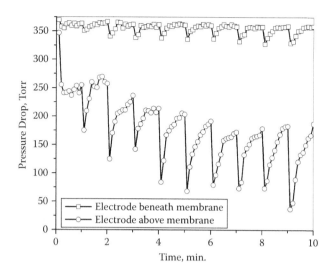

FIGURE 6.7

Comparison of pressure drop across filter membrane under repeated pulsed spark discharges in water: electrode beneath membrane versus electrode above membrane. (From Yang, Kim, et al., 2009b.)

6.3 Calcium Carbonate Precipitation with Spark Discharge

As mentioned in the introduction section, thermoelectric generation accounted for 39% (136 BGD) of all freshwater withdrawals in the United States in 2000, second only to irrigation (Feeley et al., 2008). Each kilowatt-hour (kWh) of thermoelectric generation requires the withdrawal of approximately 25 gallons of water, which is primarily consumed for cooling purposes.

Since heat removal from condenser tubes requires the evaporation of pure water, the concentration of the mineral ions, such as calcium and magnesium, in the circulating cooling water increases with time. Even though the makeup water is relatively soft, the continuous circulation eventually increases the hardness of the water due to pure water evaporation. These mineral ions, when transported through piping in an ordinary plumbing system, can cause various problems, including the loss of heat transfer efficiencies in condensers and pipe clogging due to scale formation (Panchal and Knudsen, 1998; Somerscales, 1990). Thus, to maintain a certain calcium hardness level in the cooling water, one must discharge a fraction of water through blowdown and replace it with makeup water.

In a typical cooling tower application, the COC in cooling water is often maintained at 3.5. That means if the calcium carbonate hardness of the makeup water is 100 mg/L, the hardness in the circulating cooling water is maintained at approximately 350 mg/L. If the COC can be increased through continuous precipitation and removal of calcium ions, one can significantly reduce the amounts of makeup and blowdown water, resulting in the conservation of freshwater.

Various chemical and nonchemical methods are used to prevent scaling and thus increase the COC. Among them, the scale-inhibiting chemicals like chlorine and brominated compounds were the best choice for the control of mineral fouling. Although it had a high success rate, there were also many disadvantages and concerns in their use. Aside from the high cost of chemicals, more stringent environmental laws increased the costs associated with their storage, handling, and disposal. These chemicals also pose danger to human health and the environment with accidental spills or accumulated chemical residues over a long period of time (Panchal and Knudsen, 1998; Muller-Steinhagen, 2000). Thus, there is a need for a new approach that is safe and economical from both environmental and cost points of view in cleaning and maintenance of heat exchangers. Physical water treatment (PWT) is a nonchemical method to mitigate mineral fouling with the use of electric or magnetic fields, catalytic surfaces, ultrasounds, or sudden pressure changes. Numerous studies have been reported for the effectiveness of ultrasound (Radler and Ousko-Oberhoffer, 2005); solenoid coils (Cho et al., 2004; Cho, Lane, and Kim, 2005; Kim, 2001; Ushakov, 2008; Cho, Lee, and Kim, 2003); magnetic fields (Parsons et al., 1997; Xiaokai et al., 2005; Fathi et al., 2006; Xiaokai, 2008); catalytic material (Coetzee et al., 1998; Smith et al., 2003; Lee et al., 2006); and electrolysis (Gabrielli et al., 2006; Quan et al., 2009).

Herein, the pulsed-spark-discharge-assisted precipitation of dissolved calcium ions in a hard water system is presented. First, the chemistry behind the $CaCO_3$ precipitation is described. The main reactions of precipitation and dissolution of calcium carbonate in a hard water system are (Snoeyink and Jenkins, 1982)

$$Ca^{2+}_{(aq)} + CO_3^{2-}_{(aq)} \Leftrightarrow CaCO_{3(s)} \tag{6.5}$$

and

$$CaCO_{3(s)} + H^+_{(aq)} \Leftrightarrow Ca^{2+}_{(aq)} + HCO_{3(aq)}^- \tag{6.6}$$

In a saturated condition, the forward reaction of the precipitation of $CaCO_3$ does not take place as both calcium and bicarbonate ions are hydrated. When water is supersaturated and there is sufficient energy supplied, the water molecules are disturbed or become freed from the ions, resulting in the precipitation of $CaCO_3$. Equation 6.6 shows the dissolution of solid calcium carbonate by acid, a process that takes place during acid cleaning of scaled heat exchangers.

Generally, one needs to be concerned with these two reactions. In reality, precipitation and dissociation reactions are much more complicated. The rate of recombination and crystallization of calcium and carbonate ions is controlled by three reactions. Reaction 1 relates to the dissociation of bicarbonate ions into the hydroxyl ions OH⁻ and carbon dioxide (Muller-Steinhagen, 1999):

$$HCO_{3(aq)}^- \Leftrightarrow OH^-_{(aq)} + CO_{2(aq)} \text{ (Reaction 1)} \tag{6.7}$$

The forward reaction indicates the dissociation of the bicarbonate ions, which is the most critical step from the precipitation process. Of note is that the bicarbonate ions do not cause any harm in terms of scaling as long as they remain bicarbonate ions. Reaction 1 shows the first step in the conversion of bicarbonate to carbonate ions. The presence of hydroxide ions is best indicated by a local increase in pH, and carbon dioxide typically evolves from the water as gas over time.

In reaction 2, hydroxyl ions produced from reaction 1 further react with existing bicarbonate ions, producing carbonate ions and water (Muller-Steinhagen, 1999):

$$HCO_{3(aq)}^- + OH^-_{(aq)} \Leftrightarrow H_2O_{(l)} + CO_{3(aq)}^{2-} \text{ (Reaction 2)} \qquad (6.8)$$

Reaction 3 is the reaction between calcium and carbonate ions, resulting in the precipitation and crystallization of calcium carbonate particles (Muller-Steinhagen, 1999):

$$Ca^{2+}_{(aq)} + CO_3^{2-}{}_{(aq)} \Leftrightarrow CaCO_{3(s)} \text{ (Reaction 3)} \qquad (6.9)$$

Table 6.1 presents the thermochemistry of these three reactions (Snoeyink and Jenkins, 1982). The values of DH and DG give some useful insights into the behavior of the system of reactions. The endothermic reaction 1 is the most rate limiting since it needs a relatively large input of energy to continue in the forward direction based on the high enthalpy value. The Gibbs free energy is relatively high; thus, this reaction will tend to form bicarbonate ions unless this energy restriction is overcome. For hydroxide ions to be produced as a result of dissociation of the bicarbonate ions, the energy required for the reaction can be calculated as

$$\frac{48,260J}{mol} \frac{1mol}{6.02 \times 10^{23} \, icons} = 0.801 \times 10^{-19} J \left(or \sim 0.5eV\right) \qquad (6.10)$$

To overcome the barrier, enough energy needs to be added to water so that the bicarbonate ions can be dissociated and subsequently precipitating dissolved calcium ions to $CaCO_3$ via reactions 2 and 3.

TABLE 6.1

Thermochemical Data of Reactions 1, 2, and 3

	Reaction 1	Reaction 2	Reaction 3
DH (kJ/mol)	48.26	-41.06	12.36
DG (kJ/mol)	43.60	-20.9	-47.70

Source: Snoeyink, V.L., and Jenkins, D. (1982) *Water chemistry*, Wiley, New York.

Thus, it is clear that dissociation of bicarbonate ions plays an important role in the precipitation process, as observed in the experiments. From thermodynamics, the reaction rate coefficient k for reaction 1 is

$$k = Ae^{-E_a/T} \tag{6.11}$$

where A is a constant, E_a is activation energy for the reaction, T is the system temperature (in electron volts). It is obvious that a higher temperature will lead to a higher precipitation rate. For example, when hard water is heated (i.e., volume heating) by boiling or microwave, the calcium carbonate deposit can usually be observed in the bulk solution or heated surface.

For plasma treatment cases, the total input energy was on the level of kilojoules per liter, so no significant volume heating exists. However, direct pulsed spark discharges in water have been shown to generate a temperature up to 5,000–10,000 K (about 0.5–1 eV) inside the plasma channel (Sunka et al., 1999), a phenomenon that is sufficient to induce direct pyrolysis of bicarbonate ions. The high-temperature zone is highly localized around the plasma channel, so that higher efficiency could be achieved over volume heating under the same energy input level due to an exponential dependence of the reaction rate on the temperature. The maximum temperature depends on both plasma power and water properties.

Furthermore, the thermal dissociation of bicarbonate ions may be enhanced by the emission of ultraviolet (UV) light from the high-temperature plasma channel, which functions as a blackbody radiation source. Full-spectrum UV radiation was found to be produced from a spark discharge (Sunka, 2001). The vacuum UV (VUV) ($1 = 75$–185 nm) emitted would be absorbed by the water layer immediately surrounding the plasma channel, leading to the expansion of the high-temperature zone (Locke, Sato, et al., 2006). However, the amount of solution that can be treated directly by the thermal process is still limited by the small volume of the high-temperature zone.

Another factor that may contribute to the dissociation of bicarbonate ions is the electric field. There are reports about the increase in the nucleation rate of different crystals when subjected to external electric fields (He and Hopke, 1993; Chibowski, Holysz, L., and Wojcik, 1994; Katz, Fisk, and Chakarov, 1994). Dhanasekaran and Ramasamy (1986) studied the free energy required for the formation of a two-dimensional nucleus of any possible shape under electric field. They calculated the critical free energy for nucleation as

$$\Delta G = \frac{\beta^2 \sigma^2}{4kT\left(\ln a + \varphi E^2\right)} \tag{6.12}$$

where β is a constant depending on the geometrical shape of the nucleus, σ is the interfacial tension, k is the Boltzmann constant, T is the temperature, α is the supersaturation ratio, E is the external electric field, and φ is defined as $-\varepsilon_0 v[(1/\varepsilon_g - 1/\varepsilon_l)\sin^2\theta + (\varepsilon_g - \varepsilon_l)\cos^2\theta]/8\pi kT$, where v is the volume of the crystalline nucleus, q is the angle between the direction of the electric field and the nucleation surface, and e_s and e_l are the dielectric constants of the nucleus and solution, respectively. According to Dhanasekaran's theory, the free energy for nucleation decreases as the strength of electric field increases at certain angles, which leads to a higher nucleation rate.

The effect of electric field could also be explained by the disruption of the electric double layer of hydrated ions. Dissolved calcium and bicarbonate ions do not react at room temperature as both ions are surrounded by water molecules, forming electric double layers. A number of researchers postulated that the magnetic or electric field, if strong enough, might disrupt the electric double layer and initiate the precipitation (Gehr et al., 1995; Baker and Judd, 1996; Higashitani and Oshitan, 1997). For an electric field to affect the electric double layer near a negatively charged surface directly, one needs an electric field that provides force that is able to overcome the force in the electric double layer. Typically, there are approximately 2 eV across an electric double layer, and the Debye length for a dilute solution like water is about 10 nm (Levine, 1978; Probstein, 1989). Thus, the electric field required to directly affect the electric double layer becomes approximately 2 V/10 nm $= 2 \times 10^8$ V/m. To initiate pulsed electric discharges in water using a point-to-plane electrode system, it is necessary to have a high-intensity electric field at the tip of the point electrode, which can be calculated as

$$E \approx \frac{V}{r_e} \tag{6.13}$$

where V is the applied voltage, and r_e is the radius of curvature at the tip of the electrode. For the present study, $V = 24$ kV, and $r_e \approx 200$ mm, leading to an electric field of 1.2×10^8 V/m, which is comparable to the electric field required to disrupt the hydration shells of the ions and could possibly accelerate the dehydration process of ionic pair association. In addition, unlike the highly localized thermal effect, the electric effect is not limited to the vicinity of the electrode tip. After the initiation of pulsed discharge, multichannel streamers with lengths up to several centimeters could be produced when propagating from one electrode to the other electrode. An and his coworkers (An, Baumung, and Bluhm, 2007) measured the radius of the streamers and inferred that the electric field at the tip of the self-propagating streamers was more than 2×10^9 V/m. Subsequently, the bicarbonate ions may be dissociated along the propagation path of the streamers, leading to the precipitation of calcium carbonate in bulk water as described.

6.3.1 Effect of Plasma on Cooling Water

Figure 6.8 presents the variation of pH and calcium ion concentration as a function of time and input plasma energy and their comparisons with no treatment cases. For the case of sample 1, the calcium ion concentration dropped from the initial value of 96 to 71 mg/L after 10 min of plasma

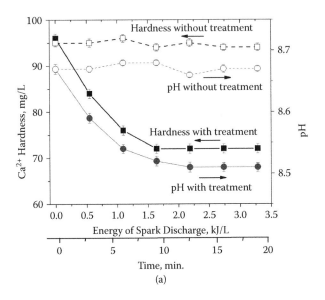

(a)

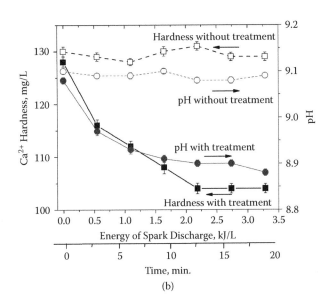

(b)

FIGURE 6.8

Variations of pH and Ca²⁺ hardness over time with and without plasma treatment: (a) Sample 1; (b) sample 2; (c) sample 3. See Table 6.2 for more information on the three samples. (From Yang, Y., Kim, H., Starikovskiy, A., Fridman, A., and Cho, Y.I. (2010) Application of pulsed spark discharge for calcium carbonate precipitation in hard water. *Water Res.* 44, 3659–3668.)

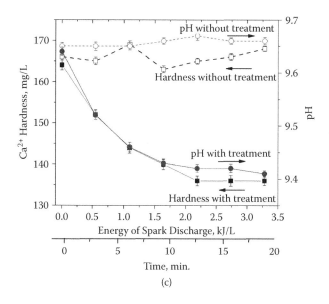

FIGURE 6.8 (CONTINUED)
Variations of pH and Ca^{2+} hardness over time with and without plasma treatment: (a) Sample 1; (b) sample 2; (c) sample 3. See Table 6.2 for more information on the three samples. (From Yang, Y., Kim, H., Starikovskiy, A., Fridman, A., and Cho, Y.I. (2010) Application of pulsed spark discharge for calcium carbonate precipitation in hard water. *Water Res.* 44, 3659–3668.)

treatment. After that, the hardness would reach an asymptotic value and not decrease with further input of energy. Accompanying the drop of the calcium ion concentration, the pH of the treated water sample decreased from 8.67 to 8.51, possibly because of the liberation of H^+ ions according to the following reaction:

$$Ca^{2+} + HCO_3^- \rightarrow CaCO_3 + H^+ \qquad (6.14)$$

Also, the ionization of water molecules by high-energy electrons from plasma discharge may contribute to the decrease of the pH through the following reaction:

$$H_2O + e \rightarrow H_2O^+ + 2e \qquad (6.15)$$

$$H_2O^+ + H_2O \rightarrow H_3O^+ + OH^* \qquad (6.16)$$

Samples 2 and 3, with initial calcium ion concentrations of 128 and 164 mg/L, respectively, showed a similar trend as sample 1. The hardness was reduced by about 25% after a 10-min treatment, with an energy input of approximately 1,800 J/L. In comparison to the plasma-treated cases, no significant change was observed for the no treatment cases.

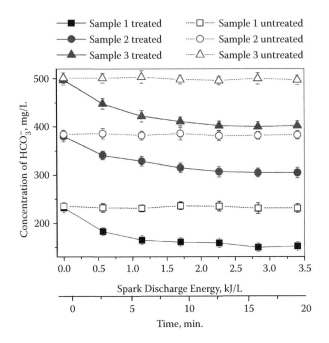

FIGURE 6.9
(See color insert.) Variations of HCO$_3^-$ over time for cases with and without plasma treatment. (From Yang, Y., Kim, H., Starikovskiy, A., Fridman, A., and Cho, Y.I. (2010) Application of pulsed spark discharge for calcium carbonate precipitation in hard water. *Water Res.* 44, 3659–3668.)

Figure 6.9 presents the variations of bicarbonate ion concentration with time determined by total alkalinity test and phenolphthalein alkalinity test. Generally, the concentration of bicarbonate ion decreased during a 10-min stabilization period before reaching an asymptotic value. The concentration of both calcium ions and bicarbonate ions changed little for energy input greater than 2 kJ/L. This was probably because the rate of precipitation reactions was limited by the diffusion rate of calcium and bicarbonate ions toward the reaction zone due to the high localization of the plasma discharge. In addition, the acidification in the vicinity of the anode by plasma-induced H$^+$ generation (Equations 6.15 and 6.16) could become detrimental to the precipitation process.

Figure 6.10 shows the number of particles per unit volume of 1 mL as a function of particle size before and after plasma treatment, which were determined using a dynamic laser-scattering system. The results depict in general that before the treatment, the number of particles less than 10 mm was significantly greater than that of larger particles (greater than 10 mm). For example, for sample 1 (Figure 6.10a), the number of particles with size between 1 and 2 mm was 17,048 before treatment as compared to 2,445 for particles greater than 10 mm. After treatment, a significant increase in the number of particles was observed for all cases as compared to the no treatment cases. For

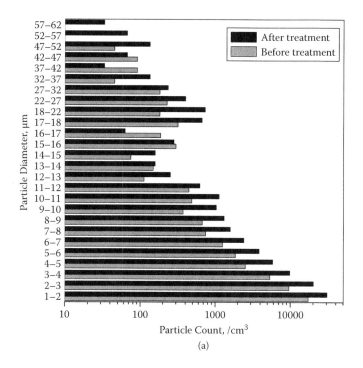

FIGURE 6.10
Particle size distributions before and after plasma treatment for (a) sample 1; (b) sample 2; (c) sample 3. (From Yang, Y., Kim, H., Starikovskiy, A., Fridman, A., and Cho, Y.I. (2010) Application of pulsed spark discharge for calcium carbonate precipitation in hard water. *Water Res.* 44, 3659–3668.)

sample 1, the number of particles with size between 1 and 10 mm increased from 39,904 at the initial state to 77,680 at the end of the treatment, while the number of particles with size over 10 mm increased from 2,445 to 3,529. For samples 2 and 3 (Figures 6.10b and 6.10c), particles smaller than 10 mm also made up the majority of the suspended solids in water. The number of small particles (i.e., below 10 mm) after the treatment was significantly increased compared with that obtained for no treatment cases.

Assuming that all the particles were in the spherical shape for the purpose of mathematical estimation, the total mass of suspended solid contents m was calculated using the following equation:

$$m = \sum_{N_d(d=1\mu m)}^{N_d(d=62\mu m)} \rho \cdot \frac{\pi d^3}{6} \cdot N_d \tag{6.17}$$

where r is the density of the $CaCO_3$ particle, d is the diameter of the particle, and N_d is the number of particles. r is taken as 2.7 g/cm³, which is the density of calcite. For sample 1, the total mass of solid particles before the treatment

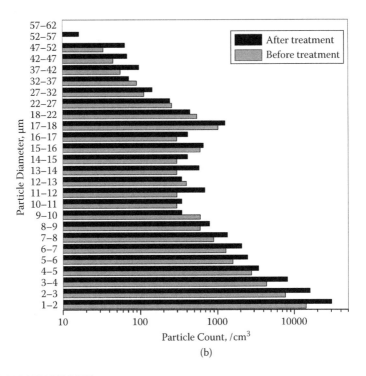

(b)

FIGURE 6.10 (CONTINUED)

Particle size distributions before and after plasma treatment for (a) sample 1; (b) sample 2; (c) sample 3. (From Yang, Y., Kim, H., Starikovskiy, A., Fridman, A., and Cho, Y.I. (2010) Application of pulsed spark discharge for calcium carbonate precipitation in hard water. *Water Res.* 44, 3659–3668.)

was 64 mg/L, which corresponded to 64 mg/L $CaCO_3$ hardness. The value increased to 104 mg/L after the plasma treatment, which means that the difference of 40 mg/L of ionic content in water was transformed from the dissolved ionic states into solid content during the process. The aforementioned titration results from Figure 6.9a showed that the calcium ion hardness in water was reduced by 25 mg/L, equaling 62.5 mg/L of calcium carbonate hardness. Considering that the resolution of the laser particle counter was 1 mm, which means that the precipitation of particles with a diameter less than 1 mm was not taken into account, the results obtained by titration and laser particle counting were in good agreement, demonstrating that the calcium ions were transformed into calcium carbonate solids during the process. For all three cases, the results are summarized in Table 6.2.

Figure 6.11a shows the scanning electron microscopic (SEM) photographs of particles retrieved from sample 1 without plasma treatment. The chemical composition of the particles was analyzed by energy dispersion spectrometry (EDS) (Figure 6.12). Figure 6.13 shows that the particles were mainly composed of calcium carbonate, with slight amounts of sodium and magnesium,

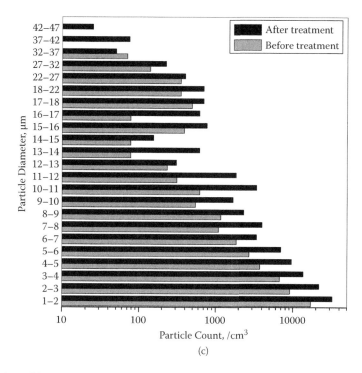

(c)

FIGURE 6.10 (CONTINUED)
Particle size distributions before and after plasma treatment for (a) sample 1; (b) sample 2; (c) sample 3. (From Yang, Y., Kim, H., Starikovskiy, A., Fridman, A., and Cho, Y.I. (2010) Application of pulsed spark discharge for calcium carbonate precipitation in hard water. *Water Res.* 44, 3659–3668.)

TABLE 6.2

Results Obtained by Laser Particle Counting

	Sample 1	Sample 2	Sample 3
Total number of particles before treatment (cm⁻³)	42,349	38,587	47,034
Mass of suspended particles before treatment	64	72	53
Total number of particles after treatment (cm⁻³)	81,710	71,340	105,846
Mass of suspended particles after treatment	104	120	108
Increase of solid mass in water	40	48	56
Decrease of ionic CaCO₃ hardness in water	62	67	75

Note: All mass values are in milligrams per liter.

FIGURE 6.11
SEM images of calcium carbonate crystals obtained from (a) untreated water; (b) plasma-treated water. (From Yang, Y., Kim, H., Starikovskiy, A., Fridman, A., and Cho, Y.I. (2010) Application of pulsed spark discharge for calcium carbonate precipitation in hard water. *Water Res.* 44, 3659–3668.)

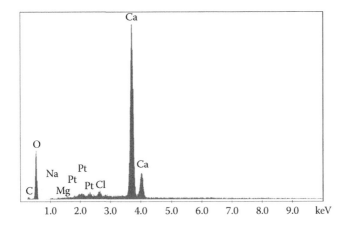

FIGURE 6.12
Elemental composition of the particles shown in Figure 6.11b obtained by energy dispersion spectrometer (EDS). (From Yang, Y., Kim, H., Starikovskiy, A., Fridman, A., and Cho, Y.I. (2010) Application of pulsed spark discharge for calcium carbonate precipitation in hard water. *Water Res.* 44, 3659–3668.)

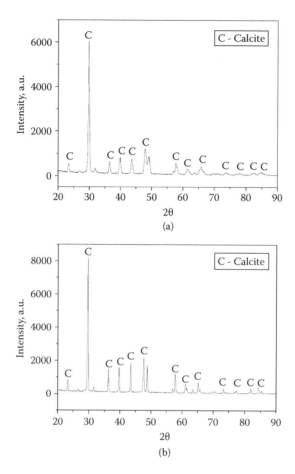

FIGURE 6.13

XRD pattern of the calcium carbonate crystals obtained from (a) untreated water; (b) plasma-treated water. (From Yang, Y., Kim, H., Starikovskiy, A., Fridman, A., and Cho, Y.I. (2010) Application of pulsed spark discharge for calcium carbonate precipitation in hard water. *Water Res.* 44, 3659–3668.)

together with the platinum coating prior to the SEM observation. The crystals exhibited morphology close to the rounded vaterite form. However, the X-ray diffraction (XRD) pattern, as shown in Figure 6.13a, coincided with that of calcite. This was probably because of the thermodynamically unstable property of the vaterite crystals. The preparation of water samples usually took several days. Possibly during this period, vaterite may have recrystallized into more stable calcite crystals without changing shape.

Figure 6.11b shows the photographs of the particles collected from the sample 1 after the plasma treatment. The crystals demonstrated the typical rhombohedron morphology of calcite. The formation of the calcite was confirmed by XRD shown in Figure 6.13b. The mean crystal size of the precipitation test was about 5 mm after 20 min of plasma treatment. For the treated water with high hardness (i.e., samples 2 and 3), calcite crystals were also observed, and their size remained almost the same as that observed in sample 1. The fact that the total amount of precipitates increased as indicated by the laser particle counting suggests that the number of nuclei was significantly increased. That demonstrates that the plasma had induced chemical changes in the calcareous water during the treatment, which was implied later in the nucleation process during the precipitation test.

6.3.2 Effect of Spray Circulation on Hardness of Cooling Water

To disturb or eliminate the acidic condition so that a higher level of hardness reduction could be achieved than the aforementioned asymptotic hardness, the treated water sample was spray circulated for 12 h to degas the excessive CO_2 in the plasma-treated water. For sample 1, the pH value increased from 7.74 to 7.92 during this period, indicating releasing of CO_2 gas from water solution. In the meantime, the hardness decreased from 190 to 160 mg/L (see two arrows in Figure 6.14a). After that, the water sample was treated again by spark discharge for 20 min, and the hardness dropped from 160 to 140 mg/L before reaching another equilibrium. At the end of a 36-h test, the hardness reached the final asymptotic value shown in Figure 6.14a, which indicated an approximate 45% reduction from the initial hardness. Similar results were observed for both sample 2 and sample 3, with hardness reductions of 53% and 59%, respectively. For example, for sample 3 with an initial hardness of 420 ppm, the final hardness was 170 ppm at the end of 36-h circulation with three brief daily plasma treatments. Note that when the plasma treatment was not used, the hardness of sample 3 increased from 420 to 440 ppm after 36-h circulation, with a 5% increase in hardness obviously due to the evaporation of pure water as shown in Figure 6.14a.

In summary, the hardness of water could be reduced by 45–59% by the combination of plasma discharges and spray circulation. Since the plasma treatment was applied for only 20 min in the 12-h interval, and spray circulation is a standard operation for cooling water in a cooling tower, plasma discharge can be regarded as a "catalyst" for precipitation.

6.3.3 Mechanism of Plasma-Induced Calcium Precipitation

In this section, the possible mechanism for plasma-induced calcium-carbonate precipitation is discussed.

6.3.3.1 Effect of Electrolysis

First, the effect of electrolysis, which is another hard-water-softening technology using electric energy, is examined. Gabrielli and his coworkers (2006) studied the principle of the softening process by electrolysis and showed

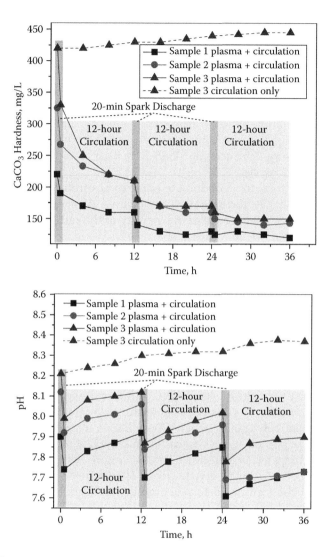

FIGURE 6.14
(See color insert.) Variations of (a) CaCO$_3$ hardness and (b) pH value over time with plasma treatment and spray circulation.

that the precipitation was induced by the local pH increase near the vicinity of the cathode. A model electrolyzer with an electrode area of 560 cm^2 was tested. Under the operating current density of 20 A/m^2, applied voltage of 12 V, and flow rate of 6 L/h, the Ca^{2+} hardness of the water sample was reduced from 96 to about 72 mg/L when a quasi-steady-state operation was achieved. Hence, the energy consumption was calculated as 806 J/L.

For the present spark discharge, electrolysis most likely took place during the prebreakdown stage as the current would be transferred by the ionized gas after the formation of the conductive channel between the two electrodes. The same level of hardness reduction was achieved as in Gabrielli's experiments after applying spark discharge in sample 1 for 10 min, as shown in Figure 6.9a. The energy dissipation in the prebreakdown stage was calculated as 984 J/L based on the results shown in Figure 6.8. However, it was difficult to estimate the exact amount of energy consumed by electrolysis as multiple physical and chemical processes, including ionization, light emission, cavitation, shock wave, and reactive species formation, were initiated during the same stage. But certainly, only a fraction of the 984 J/L energy was dissipated in electrolysis.

Another factor that may subdue the effect of electrolysis is the mixing of the liquid caused by the formation of shock waves. It may prohibit the formation of the local high pH area near the vicinity of the cathode. Therefore, the electrolysis-induced precipitation might have been suppressed in the present experiment. Hence, it could be concluded that the reduction of water hardness may not be the sole effect of the electrolysis, although it may contribute partially to the precipitation process.

6.3.3.2 Effect of UV Radiation

First, the effect of UV radiation on the precipitation process of calcium ions was studied. For spark discharges, the high-temperature plasma channel can function as a blackbody radiation source. The maximum emittance is in the ultraviolet A (UVA) to ultraviolet C (UVC) range of the spectrum produced by the spark discharge (200–400 nm), as determined by the Stephen-Boltzmann law. Water is relatively transparent to UV radiation in this wavelength range. The energy per photon ranges from 3.1 to 6.2 eV, indicating the possibility of HCO_3^- dissociation through UV absorption. Severe mineral fouling was usually observed at the quartz sleeve of commercial UV lamps, where the buildup of the fouling material could be associated with the reactions between dissolved ions and UV radiation, although the mechanism was not fully understood (Lin, Johnston, and Blatchley, 1999; Wait, 2005).

To investigate this mechanism of the reaction between dissolved mineral ions and UV radiation, the present study constructed a special discharge chamber in which a point-to-plane electrode system was placed in a quartz sleeve filled with distilled water, as shown in Figure 6.15. The diameter of the quartz sleeve was 25.4 mm to avoid possible damage by shock waves produced by the spark discharge. The quartz sleeve was found to provide a good UV window for the water samples while effectively trapping various chemical reactive species, heat, and other effects. Before each test, the quartz sleeve was cleaned using a 0.1*N* HCl solution to remove any possible $CaCO_3$ deposit from previous experiments. Test water samples were circulated outside the quartz sleeve and treated by the spark discharge produced inside the quartz sleeve for the same time periods as in the previous experiments. The results were compared to those obtained for the cases without

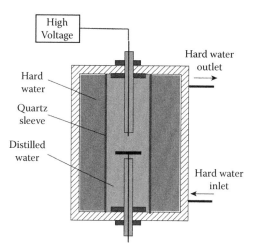

FIGURE 6.15
Modified discharge chamber to test the effect of UV radiation on the precipitation of $CaCO_3$.

the quartz sleeve, and no hardness reduction was observed for the samples treated by plasma separated by the quartz sleeve. This indicates that the process of calcium ion precipitation was not triggered by the transfer of energy to the water sample through UV radiation.

6.3.3.3 Effect of Reactive Species

From the emission spectrum of the pulsed discharge in distilled water, the formation of hydroxyl radicals and other reactive species by an underwater plasma process was reported (Sunka et al., 1999; Joshi et al., 1995; Sun et al., 1998). In the case of hard water, OH radicals would react with bicarbonate ions to produce carbonate radicals CO_3^{*-} through the following reaction (Buxton and Elliot, 1986; Crittenden et al., 1999):

$$HCO_{3(aq)}^- + OH_{(aq)}^* \Leftrightarrow H_2O + CO_{3(aq)}^{*-} \left(k = 8.5 \times 10^6 \, M^{-1} s^{-1} \right) \qquad (6.18)$$

Being a highly oxidizing species, is known to be active in the oxidation of some organic compounds by direct electron transfer to produce the corresponding cation radical and carbonate anion (Denisov et al., 2003; Crittenden et al., 1999), as shown by the following equation:

$$CO_{3(aq)}^{*-} + R \Leftrightarrow R^{*+} + CO_{3(aq)}^{2-} \qquad (6.19)$$

Hence, it is hypothesized that the reactive species produced by the spark discharge in hard water may transform the bicarbonate ions into carbonate ions without going through Equations 6.7 and 6.8. To test this hypothesis, multiple layers of polyethylene meshes were placed close to the electrodes.

Polyethylene is one of the most commonly used materials for active species scavengers in water. Typical rates of the reaction between the OH radical and organic materials are on the order of 10^9 to 10^{10} M^{-1}s^{-1}, about two orders of magnitude higher than that of the reaction between OH radicals and bicarbonate ions. Thus, the polyethylene mesh could serve as an effective scavenger screen for the radicals. The water samples were circulated and treated for the same time period with all the other experimental parameters remaining the same as in the previous experiments. No significant difference was observed in the calcium ion concentration as compared to the results shown in Figure 6.9. Therefore, it was also concluded that the reactive species were not responsible for the plasma-induced calcium carbonate precipitation.

6.3.3.4 Effect of Microheating

To gain further insight into the structure of the precipitated calcium carbonate particles, morphological and crystallization examinations were performed by SEM and XRD, respectively. Figure 6.12 shows the particles retrieved from the untreated water sample. The crystals exhibited morphology similar to a round-shaped vaterite, although XRD data showed that the particles were in a calcite form (not shown). Figure 6.5b shows the particles retrieved from the plasma-treated water sample. The crystals demonstrated a typical rhombohedron morphology of calcite, suggesting different mechanisms of precipitation after plasma treatment. Natural calcium carbonate precipitation is a complex process that has been studied extensively. It is known that calcium carbonate exists in three crystalline polymorphs with different structures: calcite, aragonite, and vaterite. Among them, calcite is the most thermally stable form and is the dominant polymorph of $CaCO_3$ formed by the loss of carbon dioxide or evaporation of natural calcium bicarbonate solutions if temperature is the controlling factor (Siegal and Reams, 1966). The other two crystalline forms are metastable phases of calcium carbonate, which would transform to calcite spontaneously under normal conditions. The transformation process would be expedited on heating. Therefore, from the fact that calcite with a rhombohedron morphology was formed during the plasma-assisted precipitation process, one may hypothesize that the precipitation may be temperature related, i.e., through localized microheating from the plasma charge.

Direct proof of the validity of this hypothesis is difficult as it is not easy to produce a local heating zone with temperature up to 5,000–10,000 K in water without inducing other effects. To test this hypothesis, the present study utilized a transient hot-wire method. When a thin wire immersed in a sample liquid is heated by electrical current (i.e., joule heating), the wire can become an electrical heating element and produce an elevated temperature in the surrounding water. The transient hot-wire technique is widely employed today for the measurement of the thermal conductivity of fluids over a wide range of temperatures (Nagasaka and Nagashima, 1981; Xie et al., 2006). Although the temperature rise in the conductivity measurement is usually much lower than that in the spark discharge channel, the hot-wire method is still a good approximation as the duration of the high temperature produced by the spark discharge is relatively short (usually in microseconds), and the intense local heating would quickly be dissipated through the conductive and convective heat transfer in water.

A platinum wire 80 mm in diameter and 20 mm long was used in the present study. The resistance of the effective length of the wire was about 10 Ω at 25°C. The power for the circuit for heating the wire was provided by a square-wave alternating voltage unit, with the voltage waveform shown in Figure 6.16. The wire was heated during a duty cycle and cooled during an off-duty cycle when $V = 0$. The temperature rise of the wire ΔT can be given by the following equation (Assael et al., 1998):

$$\Delta T = \left(\frac{q}{4\pi\lambda}\right) In\left(\frac{4\kappa t}{a^2 C}\right) \tag{6.20}$$

where λ and k are the thermal conductivity and diffusivity of the liquid, respectively; q is the heat generation per unit length of the wire; a is the radius of the wire; $C = e^y \approx 1.781$; y is Euler's constant; and t is the time of heating. With $R = 10 \Omega$ and $l = 20$ mm, q can be calculated as

$$q = \frac{\frac{V^2}{R}}{l} = \frac{2.5 \times 10^4 \, W}{m} \tag{6.21}$$

Hence, the temperature rise at the end of the heating period was about 825 K. The actual temperature rise would be much lower than the theoretical

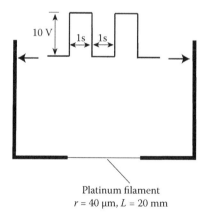

Platinum filament
$r = 40$ μm, $L = 20$ mm

FIGURE 6.16
Schematic diagram and voltage waveform used in the transient hot-wire method. (From Yang, Y., Kim, H., Starikovskiy, A., Fridman, A., and Cho, Y.I. (2011b) Precipitation of calcium ions from hard water using pulsed spark discharges and its mechanism. *Plasma Chem. Plasma Proc.* 31, 51–66.)

value due to the boiling of water. During the cooling period, the time constant for cooling can be given as (Incropera et al., 2007)

$$\tau = \frac{\rho V c_p}{hA} \tag{6.22}$$

where ρ, V, c_p, and A are the density, volume, specific heat, and surface area of the platinum wire, respectively; and h is the convective heat transfer coefficient of water. Substituting the values for ρ, V, c_p, and A into Equation 6.22, one can obtain

$$\tau \approx \frac{100}{h} s \tag{6.23}$$

For water, the typical convective heat transfer coefficient h is 300–10,000 W/(m²K) (Incropera et al., 2007), which means under normal conditions, the time constant for cooling t is significantly less than 1 s, and the wire would be cooled to room temperature during the cooling period.

Figure 6.17 shows the effect of transient hot wire on the changes in calcium ion concentration for three different water samples. For all three cases, about a 15% hardness decrease was observed for a similar level of energy input despite the much lower temperature gradient compared to that in the plasma treatment cases. Considering the maximum 25% hardness drop in the case of the pulse spark discharge, one can conclude that the local microheating can be one of the major pathways to the precipitation of calcium ions in hard water.

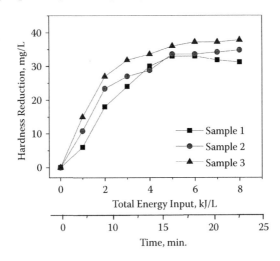

FIGURE 6.17
Calcium carbonate hardness reduction versus input energy by the transient hot-wire method. (From Yang, Y., Kim, H., Starikovskiy, A., Fridman, A., and Cho, Y.I. (2011b) Precipitation of calcium ions from hard water using pulsed spark discharges and its mechanism. *Plasma Chem. Plasma Proc.* 31, 51–66.)

FIGURE 6.18
SEM images of calcium carbonate scale obtained from hot-wire surface. (From Yang, Y., Kim, H., Starikovskiy, A., Fridman, A., and Cho, Y.I. (2011b) Precipitation of calcium ions from hard water using pulsed spark discharges and its mechanism. *Plasma Chem. Plasma Proc.* 31, 51–66.)

White deposits were observed on the filament immediately after the application of the pulsed voltage. At $t = 15$ min, the filament was fully covered by calcium carbonate scale when hardness reduction reached an asymptotic value. This could be attributed to the fact that the wire could not function as a hot surface in water anymore because of the accumulation of the thermal-insulating layer of $CaCO_3$ scale. The scale on the filament was examined using SEM (Figure 6.18) and XRD. Rhombohedral shaped calcite was observed, which was similar to the shape of calcium carbonate particles collected in the plasma treatment cases. Figure 6.19 shows the number of particles per 1-mL water sample as a function of particle size before and after the hot-wire treatment for sample 1. No significant increase in the number of particles was observed as compared to the no treatment cases, possibly because most of the calcium carbonate precipitated on the hot-wire surface instead of in bulk water.

6.3.3.5 Nonthermal Effect of Plasma

To study the influence of other products produced by plasma discharge and avoid the introduction of any thermal effects, a nanosecond pulsed power supply was constructed by adding the second spark gap in the circuit as shown in Figure 6.20. The second spark gap was built with an adjustable interelectrode distance to adjust the duration of the pulse so that one could remove the voltage from the load. The typical voltage and current waveforms are shown in Figure 6.21. The pulsed source provided 20-kV pulses with about 10 ns in duration at a repetition rate of approximately 38 Hz. Details of the corona discharge

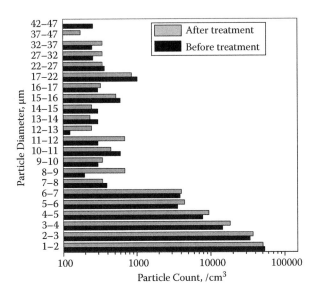

FIGURE 6.19
Particle size distributions before and after transient hot-wire treatment. (From Yang, Y., Kim, H., Starikovskiy, A., Fridman, A., and Cho, Y.I. (2011b) Precipitation of calcium ions from hard water using pulsed spark discharges and its mechanism. *Plasma Chem. Plasma Proc.* 31, 51–66.)

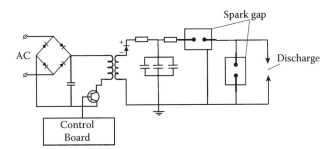

FIGURE 6.20
Schematic diagram of the double spark gap configuration. AC, alternating current.

produced were reported elsewhere (Staack et al., 2008). Spectroscopy measurement through the expansion of the Hα emission line demonstrated an almost nonthermal condition inside the plasma channel, confirming that the effect of microheating on the precipitation of calcium ions was eliminated.

Figure 6.22 shows the changes in calcium ion concentration with the application of the nanosecond corona discharge. A maximum 7% drop in the hardness was observed, demonstrating the possibility to trigger the precipitation process through nonthermal discharges. The precipitation rate decreased with time, reaching zero at $t = 20$ min. After that, the hardness value began to increase, possibly because precipitated calcium carbonate particles began to redissolve due to the ionization of water molecules and subsequent

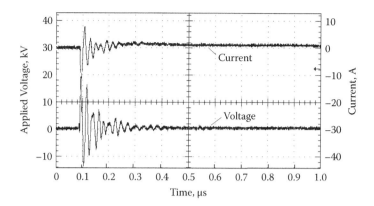

FIGURE 6.21
Typical voltage and current waveform produced by the circuit in the double spark gap configuration.

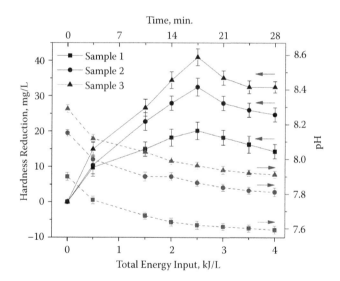

FIGURE 6.22
(See color insert.) Variations of calcium carbonate hardness and pH over time for different energy inputs by the transient hot-wire method. (From Yang, Y., Kim, H., Starikovskiy, A., Fridman, A., and Cho, Y.I. (2011b) Precipitation of calcium ions from hard water using pulsed spark discharges and its mechanism. *Plasma Chem. Plasma Proc.* 31, 51–66.)

acidification of the solution, as illustrated by Equations. 6.15 and 6.16. A drop in pH was observed for all three water samples, as shown in Figure 6.22.

Figure 6.23 shows the number of particles per 1-mL water sample as a function of particle size before and after treatment. A significant increase in the number of particles, especially for particles with diameter below 10 mm, was observed, indicating the occurrence of precipitation in bulk volume. The

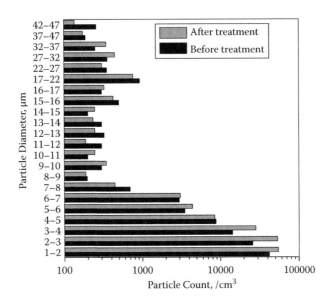

FIGURE 6.23

Particle size distributions before and after pulsed nanosecond discharge treatment. (From Yang, Y., Kim, H., Starikovskiy, A., Fridman, A., and Cho, Y.I. (2011b) Precipitation of calcium ions from hard water using pulsed spark discharges and its mechanism. *Plasma Chem. Plasma Proc.* 31, 51–66.)

FIGURE 6.24

SEM image of calcium carbonate particles obtained from water sample treated by pulsed nanosecond discharge. (From Yang, Y., Kim, H., Starikovskiy, A., Fridman, A., and Cho, Y.I. (2011b) Precipitation of calcium ions from hard water using pulsed spark discharges and its mechanism. *Plasma Chem. Plasma Proc.* 31, 51–66.)

particles suspended in the bulk volume were collected and examined using both SEM and XRD. Figure 6.24 shows a mixture of irregular and rhombohedral-shaped calcite. The former was probably preexisting in the water sample before the plasma treatment. The rhombohedral-shaped calcite was probably formed during the plasma treatment and was in a similar polymorph with calcium carbonate particles collected in the regular spark discharge treatment.

6.3.3.6 Discussions of Calcium Precipitation with Plasma

The present study proposes that local microheating is the primary mechanism in the calcium hardness reduction by spark discharges. One can easily precipitate supersaturated calcium ions in hard water by simply increasing water temperature (i.e., *volume heating*). This is technically sound and simple, but it is not a real solution due to the high cost associated with volume heating. Hence, instead of volume heating, the present study attempted to dissociate the bicarbonate ions using spark discharges in water, which is referred as *local microheating*.

For comparison, one can calculate the numbers of OH^- from both volume heating and local heating to see the benefit of the local heating in the precipitation of calcium ions. The amount of OH^- per unit time that one can produce from the dissociation reaction of bicarbonate ion, Equation 6.9, is calculated as (Fridman and Kennedy, 2006)

$$n_{OH^-} = n_{HCO_3^-} \times k \tag{6.24}$$

where $n_{HCO_3^-}$ is the number of HCO_3^- participating in the reaction, and k is the reaction rate coefficient. According to the Arrhenius equation, the reaction rate coefficient k becomes

$$k = Ae^{-E_a/T} \tag{6.25}$$

where E_a is activation energy, T is the system temperature (in electron volts). Due to the exponential curve of the equation, the Arrhenius equation indicates that *the higher the water temperature is, the faster the reaction will be.* The present study utilized spark discharges in water. Hence, one can expect intense local heating of a small volume of water around the tip of the electrode, significantly raising the temperature of the small volume of water near the tip.

The important scientific issue is whether the spark discharge used in the present study can dissociate HCO_3^- without spending a large amount of electrical energy. One can examine two cases (i.e., volume heating and local heating) to find out which case produces more OH^- for exactly the same energy spent.

For the volume heating, one can assume to heat the entire volume of water by 1 degree (e.g., from 300 to 301 K). Then, the number of OH^- one can produce for $E_a \approx 1$ eV becomes

$$n_{OH^-} = n_{HCO_3^-} \times k = n_{HCO_3^-} \times Ae^{-E_a/T} = An_{HCO_3^-}e^{-11000/301} = e^{-36.5}An_{HCO_3^-} \tag{6.26}$$

For local heating using spark discharge, one can assume heating 1% of the entire water volume by 100 degrees (e.g., from 300 to 400 K). The number of HCO_3^- participating in the reaction is 1%, that is, $n'_{HCO_3} = 0.01 \times n_{HCO_3}$, because spark discharge is assumed to heat only 1% of the total water volume. Then, the number of OH^- one can produce for $E_a \approx 1$ eV becomes

$$n'_{OH^-} = n'_{HCO_3} \times k' = n'_{HCO_3} \times Ae^{-E_a/T} = 0.01An_{HCO_3}$$

$$e^{-11000/400} = 0.01e^{-27.5}An_{HCO_3} = e^{-32}An_{HCO_3}$$

(6.27)

Comparing the number of the hydroxyl ions produced for the two cases, that is, n_{OH^-} and n'_{OH^-}, one can see that the local heating by spark discharge can produce about 100 times more OH^-, thus 100 times more efficiently precipitating dissolved calcium ions in hard water than the volume heating.

In conclusion, it has been demonstrated earlier that pulsed plasma discharge can trigger the precipitation process of calcium carbonate in hard water (Yang, Kim, Sarikovskiy, et al., 2010). The equilibrium of calcium ion concentration after approximately 10-min plasma treatment was observed in previous experiments. The possibility to shift the equilibrium and enhance the precipitation was demonstrated by degassing the dissolved CO_2 from water through spray circulation. It was hypothesized that the precipitation process was associated with different effects produced by the discharge. Experiments were conducted showing that UV radiation or reactive species produced by the spark discharge were negligible for the precipitation process. The effect of microheating was tested using a hot-wire method, while the nonthermal effect of the plasma was investigated by application of a nanosecond pulsed electric discharge in water.

It was observed that both cases showed about 10% drop of calcium ion concentration, indicating that the precipitation process may be associated with both the thermal and nonthermal effect of plasma in water. The morphology of the calcium carbonate particles collected from the two experiments was in agreement with that collected from water samples treated by conventional thermal spark discharge, indicating that the precipitation process may be associated with both the thermal and nonthermal effect of plasma. Further investigations are necessary to determine the detailed pathways of calcium carbonate precipitation by the pulsed plasma treatment.

6.3.4 Economic Analysis of Plasma Water Treatment

Finally, the anticipated benefits of the present plasma water treatment using spark discharges are discussed from the economic point of view. For a modern 1,000-MW fossil-fueled power plant with 40% efficiency, it would reject 1,500 MW of heat at the full load and use about 2,800 m^3/min of circulating water based on 10°C temperature difference in condenser (El-Wakil, 1984). As heat is removed via evaporation of pure water at a cooling tower, the need for the makeup water is about 28 m³/min to compensate the loss through evaporation, wind drift, and blowdown (El-Wakil, 1984). Assuming that the makeup water is moderately hard with a $CaCO_3$ hardness of 100 mg/L, the total amount of $CaCO_3$ that the makeup water brings into the cooling tower becomes

$$\dot{m}_{CaCO_3} = 28m^3/\min\ 100mg/L = 2800g/\min \tag{6.28}$$

In a typical cooling tower application, the COC is usually maintained at 3.5. That means the hardness in the circulating cooling water is maintained at approximately 350 mg/L, and the blowdown rate can be calculated if the wind drift loss is neglected:

$$q_{blowdown} = \frac{q_{makeup}}{COC} = 8m^3/\min \tag{6.29}$$

It has been demonstrated that the present pulsed plasma technology could continuously precipitate Ca^{2+} from water and potentially allows a cooling tower to operate at a higher COC than the current standard due to the reduction in blowdown frequency. Ideally, zero blowdown could be achieved if all the Ca^{2+} brought in by the makeup water is precipitated by the plasma; thus, the constant mineral concentration is maintained in the main cooling loop without blowdown. The energy cost, as shown, is about 1,800 J/L to achieve an average 25% reduction in water hardness. Assuming that the cooling water is treated in a side-stream loop, the flow rate needed becomes

$$q = \frac{\dot{m}_{CaCO_3}}{350mg/L \times 25\%} = 32m^3/\min \tag{6.30}$$

which is approximately 1.2% of the flow rate of the main loop. The power needed to treat the water in the side-stream loop can be calculated as

$$P = \frac{1800J/L \times 32 \times 10^3 L/\min}{60s/\min} = 960kW \tag{6.31}$$

which is 0.1% of the full capacity of the 1,000-MW power plant. Meanwhile, the flow rate of the makeup water can be reduced from 28 to 20 m³/min, which is equal to a saving of approximately 11,500 m³ per day due to the elimination of blowdown.

6.4 Application for Mineral Fouling Mitigation in Heat Exchangers

Calcium carbonate is the most common scale-forming mineral occurring in industrial water facilities. It is generally the first mineral to precipitate out either by heating or by concentrating water due to its relatively low solubility, although its concentration in source water significantly varies depending on locations. Control of calcium carbonate scale is thus often the limiting factor in most industrial cooling water applications as it decreases the efficiency of heat exchangers because of the insulating effect of the deposits. Furthermore, the formed deposits reduce the opening area in heat exchanger tubes, thus requiring more pumping power if one desires to maintain a constant flow rate (Bott, 1995; Panchal and Knudsen, 1998; Somerscales, 1990; Cho, Lane, and Kim, 2005; Cho, Lee, and Kim, 2003). An 0.8-mm layer of $CaCO_3$ scale can increase the energy use by about 10% (Donaldson and Grimes, 1988; U.S. Department of Energy, 1998). If one can prevent or mitigate fouling on heat transfer surfaces, it not only increases heat exchanger efficiency, but also reduces the expenses associated with the cleaning of fouled heat exchangers. In addition to the benefit of reduced mineral fouling, the COC can be increased, resulting in water savings by reduced makeup and blowdown as shown in the previous section and other previous studies (Muller-Steinhagen, 2000; Radler and Ousko-Oberhoffer, 2005; Tijing et al., 2009; Demadis et al., 2007; Cho et al., 2007). Blowdown is the water drained from cooling equipment to remove mineral buildup in the circulating water.

Various chemical and nonchemical methods have been used to prevent mineral fouling. Among them, the scale-inhibiting chemicals like brominated compounds were the best choices as they had a relatively high success rate. Due to safety issues and environmental concerns with the chemicals, there is a need for a new approach that is safe and clean from both environmental and cost points of view in cleaning and maintenance of heat exchangers. PWT is a nonchemical method to mitigate mineral fouling, as described previously in this chapter. Numerous studies have been reported for the effectiveness of ultrasonic, solenoid coils, magnetic fields, catalytic material, and electrolysis methods. Yang and his coworkers (Yang, Kim, Sarikovskiy, et al., 2010) reported that oversaturated hard water treated by underwater pulsed spark discharge may induce the precipitation of calcium carbonate in supersaturated water and produce a significantly greater number of particles than the untreated water. Note that the precipitation of dissolved mineral ions takes place in the bulk water instead of on the heat exchanger surfaces. This is the key process for all PWT methods as the particles suspended in water tend to form a soft coating on heat transfer surfaces. If the shear force produced by flow is large enough to remove the soft coating, mineral fouling can be prevented or mitigated.

In the present study, direct pulsed spark discharge generated in water is used to mitigate mineral fouling in a double-pipe heat exchanger. The new method of using microsecond-duration pulsed plasma in water is a major improvement over the aforementioned PWT because the previous PWT induced electric fields in water, where the field strength is often small (~1 mV/cm) due to involved physics laws, such as Faraday's law. In comparison, an electric field above 10^6 V/cm could be produced by pulsed plasma in water in the present study, leading to higher efficiency than the previous PWT devices.

The present study conducted fouling experiments in a heat exchanger by circulating artificially prepared hard water through a simulated cooling tower system. Figure 6.25 shows the schematic diagram of the present test facility, which consisted of two separate loops for circulation of hot and cooling water, a pulsed spark discharge generation system, a cooling tower, a heat exchanger test section, pumps, and a data acquisition system.

Figure 6.26 shows the schematic diagram of the counterflow concentric-type heat exchanger. The heat transfer rate Q was calculated from both hot and cooling water sides as

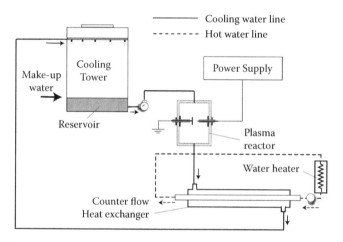

FIGURE 6.25
Schematic diagram of the experimental setup. (From Yang, Y., Kim, H., Fridman, A., and Cho, Y.I. (2010) Effect of a plasma-assisted self-cleaning filter on the performance of PWT coil for the mitigation of mineral fouling in a heat exchanger. *Int. J. Heat Mass Transfer* 53, 412–422.)

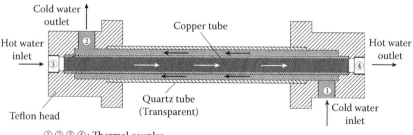

FIGURE 6.26
Schematic diagram of the heat transfer test section. (From Yang, Y., Kim, H., Fridman, A., and Cho, Y.I. (2010) Effect of a plasma-assisted self-cleaning filter on the performance of PWT coil for the mitigation of mineral fouling in a heat exchanger. *Int. J. Heat Mass Transfer* 53, 412–422.)

$$Q = \dot{m}_h c_p \Delta T_h = \dot{m}_c c_p \Delta T_c \qquad (6.32)$$

where $\dot{m}_h$ and $\dot{m}_c$ are the mass flow rates of hot and cooling water, respectively; c_p is the specific heat of water; and DT_h and DT_c are the temperature differences between inlet and outlet of hot and cooling water, respectively. The heat transfer rates at hot and cooling water sides should be equal under ideal conditions. In reality, the heat transfer rate in the hot water side was less, approximately 5%, than that in the cold water side as parasitic heat loss takes place to the surroundings in spite of insulation. Hence, the heat transfer rate measured from the cooling water side was used to calculate the overall heat transfer coefficient. The heat transfer rate Q varied from 1.9 to 3.2 kW, depending on the flow velocity at the cold water side.

The overall heat transfer coefficient U was calculated using the following equation (Incropera et al., 2007):

$$U = \frac{Q_c}{A_0 \Delta T_{LMTD}} \qquad (6.33)$$

The heat transfer surface area A_o was calculated using the outer diameter of the copper tube (d_o = 22.2 mm) with an effective heat transfer length of 600 mm (i.e., $A_o = pd_oL_{effective}$). The log mean temperature difference ΔT_{LMTD} was determined as follows (Incropera et al., 2007):

$$\Delta T_{LMTD} = \frac{\left(T_{h,o} - T_{c,i}\right) - \left(T_{h,i} - T_{c,o}\right)}{\ln\left[\frac{\left(T_{h,o} - T_{c,i}\right)}{\left(T_{h,i} - T_{c,o}\right)}\right]} \qquad (6.34)$$

The fouling resistance R_f was calculated using the following equation (Incropera et al., 2007):

$$R_f = \frac{1}{U_f} - \frac{1}{U_i} \qquad (6.35)$$

where U_f is the overall heat transfer coefficient for fouled states, while U_i is the overall heat transfer coefficient corresponding to the initial clean state. The latter (U_i) was determined using distilled water (without chemicals) and without the use of a PWT device during the initial calibration run prior to the fouling tests with artificial hard water.

6.4.1 Fouling Resistance: Validation Study

Figure 6.27 shows the results for the fouling tests obtained using a water hardness of 250 ppm for the no treatment and plasma-treated cases at a flow velocity of 0.1 m/s. Due to high water hardness, there was no induction

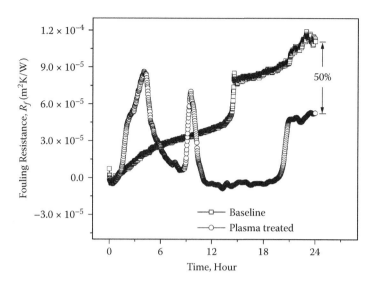

FIGURE 6.27
Fouling resistances for 250 ppm hard water under no treatment and plasma-treated cases with a flow velocity of 0.1 m/s. (From Yang, Y., Kim, H., Fridman, A., and Cho, Y.I. (2010) Effect of a plasma-assisted self-cleaning filter on the performance of PWT coil for the mitigation of mineral fouling in a heat exchanger. *Int. J. Heat Mass Transfer* 53, 412–422.)

period in both cases. An induction period is usually depicted by a straight horizontal line in the beginning of the fouling curve, which indicates lateral spreading of scale deposits on the heat transfer surface. In the present study, the artificial hard water that contained calcium and bicarbonate ions reacted quickly to the hot heat transfer surface, making immediate depositions of calcium salt particles on the surface as soon as the fouling test began.

The scale deposition involved the cumulative effect of a direct diffusion of dissolved calcium ions to the heat transfer surface and the deposition of precipitated calcium salt particles due to supersaturated conditions and accelerated precipitation of calcium salts by PWT. The fouling resistance in the no treatment case demonstrated a slow increase in the first 14 h of operation. At $t = 15$ h, the fouling resistance increased dramatically as the entire surface of copper tube was fully covered by mineral scales. After $t = 16$ h, the fouling resistance began to rise as the thickness of the scale layer slowly increased until the end of the test, indicating that the deposition rate of the scales was consistently larger than the removal rate during this period because of the slow flow velocity.

The fouling resistance curves obtained in the cases for the plasma treatment depicted a completely different trend compared to that obtained for the no treatment case. The fouling resistance had a steep increase to a maximum value in the first 4 h of operation. Note that there are two different categories of fouling: particulate fouling and precipitation fouling. The former refers to the adhesion of suspended particles to the heat transfer surface in the form of soft sludge. This type of fouling can easily be removed by shear forces created by flow compared to those deposits produced from the precipitation of mineral ions directly on the solid heat transfer surface (i.e., precipitation fouling). It was demonstrated in our previous study that the precipitation of calcium carbonate could be induced by application of pulsed spark discharge in supersaturated hard water, thus creating a significantly greater number of $CaCO_3$ particles than the untreated water. Hence, much faster particulate fouling took place during the first several hours of the test, causing the dramatic increase in the fouling resistance. At $t = 4$ h, the fouling resistance showed a significant drop, indicating that large-scale pieces were dislodged due to the shear stress of the water flow. The similar particulate fouling buildup and dislodging process were repeated during the period between 9 and 12 h. The final asymptotic fouling resistance at the end of the test was 50% lower than that obtained from the baseline test, clearly indicating the beneficial effect of the plasma discharge on the mitigation of mineral fouling.

Figure 6.28 shows photographs of sections of fouled copper tubes for the no treatment and plasma-treated cases; the photographs were taken after

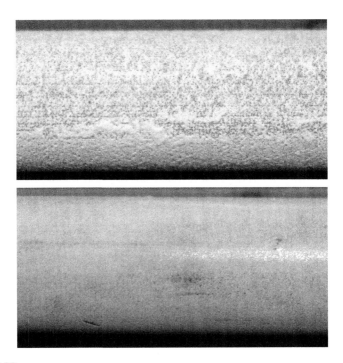

FIGURE 6.28
Photographic images of the scales for (a) no treatment and (b) plasma treatment cases ($CaCO_3$ hardness of 250 ppm and a flow velocity of 0.1 m/s). (From Yang, Y., Kim, H., Fridman, A., and Cho, Y.I. (2010) Effect of a plasma-assisted self-cleaning filter on the performance of PWT coil for the mitigation of mineral fouling in a heat exchanger. *Int. J. Heat Mass Transfer* 53, 412–422.)

the copper tubes were removed from the heat transfer test section and completely dried. Visual inspection on the fouled tubes indicated that there were thick scale (>1 mm) deposits over the entire tube surface for the no treatment case. For the case of the plasma treatment, the scale deposits appeared to be much thinner than that observed in the no treatment case. One could clearly see the copper-tone color of the tube at the end of the fouling test for the plasma-treated case, indicating that the pulsed spark discharge could significantly mitigate the scale deposits on the tube surface.

Figure 6.29 shows the results for the fouling tests obtained using a water hardness of 250 ppm for both cases at a flow velocity of 0.5 m/s with zero blowdown. The overall fouling resistance showed a significant drop compared with that obtained at 0.1 m/s, mainly because of the higher removal rate caused by a higher flow velocity. For the no treatment case, the fouling resistance had a steep increase to a local maximum value in the first 3 h of operation. At $t = 3$ h, the fouling resistance stopped increasing, and the value remained unchanged for the next 16 h, indicating that there must have been some balance between the deposition rate and removal rate for the no treatment case due to the higher shear stress produced by a high flow velocity. At $t = 19$ h, the no treatment case showed a significant drop in the fouling resistance. After $t = 30$ h, the fouling resistance showed slight up-and-down trends, with its mean value slightly decreasing with time until the end of the test.

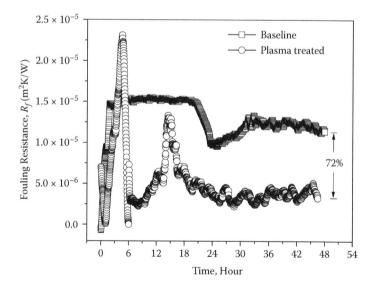

FIGURE 6.29
Fouling resistances for 250 ppm hard water under no treatment and plasma-treated cases with a flow velocity of 0.5 m/s. (From Yang, Y., Kim, H., Fridman, A., and Cho, Y.I. (2010) Effect of a plasma-assisted self-cleaning filter on the performance of PWT coil for the mitigation of mineral fouling in a heat exchanger. *Int. J. Heat Mass Transfer* 53, 412–422.)

The fouling resistance curves obtained in the case for plasma treatment at 0.5 m/s depicted a similar trend as the 0.1-m/s case shown in Figure 6.29 at the first 20-h period, although at a much lower value. The fast rise and fall of fouling resistance observed in the first 6-h period indicated both accumulation and removal of particulate fouling on the heat exchanger surface. After $t = 20$ h, the fouling resistances slightly went up and down numerous times until the end of the test. The up-and-down trends of the fouling resistance clearly indicated that the old scales were repeatedly removed from the heat transfer surface as the new scales continued to develop. The surface was not fully covered by scales at the end of the test, and the final fouling resistance was reduced by 72% compared to that for the no treatment case. The improved efficiency in mitigating fouling observed in the plasma-treated case can be explained as follows: Calcium ions were continuously precipitated to calcium salt particles by the spark discharge. Subsequently, particulate fouling took place as calcium particles adhered to the heat transfer surface, creating a soft sludge coating on the surface. The coating can be more easily removed due to high flow. Similar results were reported previously with a solenoid coil by Cho et al. (Cho et al., 2004; Cho, Lane, and Kim, 2005; Cho, Lee, and Kim, 2003), who showed better mitigation results at a high flow rate in the study.

Figures 6.30a and b show the results for the fouling tests obtained for the case of water hardness of 500 ppm and flow velocities of 0.1 and 0.5 m/s, respectively, for the plasma treated and no treatment cases. Due to the high water hardness used in the present study, the rate of increase in the fouling resistance was very steep for the no treatment case at 0.1 m/s, which reached the local maximum at $t = 3$ h. After this point, the fouling resistance significantly decreased until $t = 5$ h, indicating that the removal of scale particles was greater than the new deposits during this period, probably due to a reduced opening in the heat transfer test section by the scale deposits and subsequently increased wall shear stress. Again, the similar cycle was repeated during the 5- to 9-h period. Note that the supersaturation level in the cooling water was extremely high, so there must have been a large number of suspended $CaCO_3$ particles in the water even in the no treatment case, leading to particulate fouling on the heat transfer surface. Thus, one might expect that the scale deposits might have been soft, which helped increase the removal rate.

For the plasma treatment, there was an induction period of approximately 3 h before a sudden but brief increase of fouling resistance, indicating that the heat exchanger surface was fully covered by the scales at $t \approx 3$ h. After $t \approx 4$ h, the fouling resistance impressively remained constant until the end of the test. The asymptotic value for the plasma-treated case was about 88% lower than that for the no treatment case.

Figure 6.30b presents the fouling resistance for the flow velocity of 0.5 m/s, showing that there was also no induction period for the no treatment case. As the velocity of cold water in the heat transfer test section

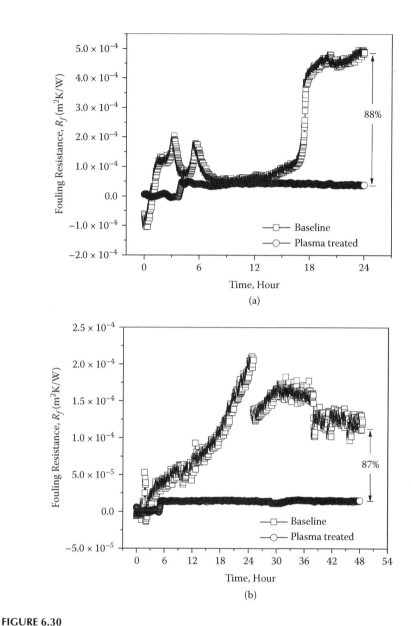

FIGURE 6.30
Fouling resistances for 500 ppm hard water under no treatment and plasma treatment cases with two different flow velocities: (a) 0.1 m/s; (b) 0.5 m/s. (From Yang, Y., Kim, H., Fridman, A., and Cho, Y.I. (2010) Effect of a plasma-assisted self-cleaning filter on the performance of PWT coil for the mitigation of mineral fouling in a heat exchanger. *Int. J. Heat Mass Transfer* 53, 412–422.)

was increased from 0.1 to 0.5 m/s, there was less fouling deposit in general as the removal rate increased due to increased shear force, a phenomenon that was also reported by another study (Yang, Gutsol, et al., 2009). Note that at a high velocity, there was a high mass deposition rate. However, the shear force created by the flow increased such that the scales were more efficiently removed, resulting in reduced fouling resistances. An 87% drop in the fouling resistance was obtained for the plasma-treated case compared with the no treatment case, again confirming the effectiveness of the plasma treatment of water on mitigating the mineral fouling. It is of note that even for the case of 500 ppm, for which the hardness became about 1,500 ppm near the end of test with zero blowdown, the plasma-treated case could reduce the fouling resistance by 88% and 87% for flow velocities of 0.1 and 0.5 m/s, respectively.

Figure 6.31 shows photographs of sections of fouled copper tubes for both no treatment and plasma-treated cases for water with a hardness of 500 ppm. The photographs taken for the 500-ppm case were similar to those obtained for the 250-ppm case, but the scale was thicker due to the higher hardness and smaller flow velocity.

6.4.2 Visualization of the Calcium Carbonate Particles

Figures 6.32a and 6.32b show SEM images of $CaCO_3$ scales for both the no treatment and plasma-treated cases for 250-ppm hard water at a flow velocity of 0.5 m/s. The SEM images for the no treatment case showed particles less than 10 mm in size, with sharp and pointed tips in crystal structures, whereas those obtained with plasma treatment showed particles 15 mm in

FIGURE 6.31
Photographic images of the scales obtained for (a) no treatment and (b) plasma treatment cases ($CaCO_3$ hardness of 500 ppm and a flow velocity of 0.1 m/s). (From Yang, Y., Kim, H., Fridman, A., and Cho, Y.I. (2010) Effect of a plasma-assisted self-cleaning filter on the performance of PWT coil for the mitigation of mineral fouling in a heat exchanger. *Int. J. Heat Mass Transfer* 53, 412–422.)

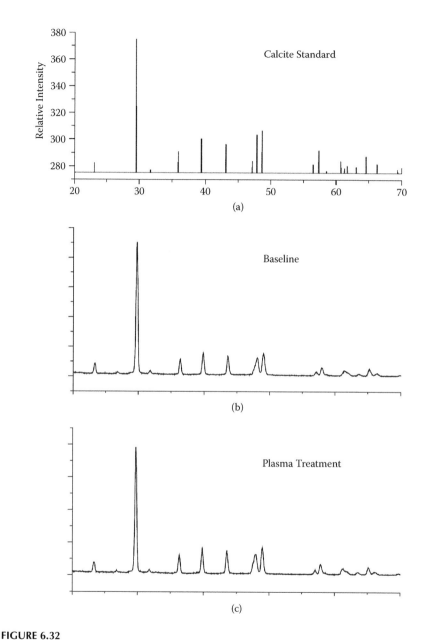

FIGURE 6.32
SEM photographs of the scales obtained for (a) no treatment and (b) plasma-treated cases
($CaCO_3$ hardness of 250 ppm and a flow velocity of 0.5 m/s). (From Yang, Y., Kim, H.,
Fridman, A., and Cho, Y.I. (2010) Effect of a plasma-assisted self-cleaning filter on the per-
formance of PWT coil for the mitigation of mineral fouling in a heat exchanger. *Int. J. Heat
Mass Transfer* 53, 412–422.)

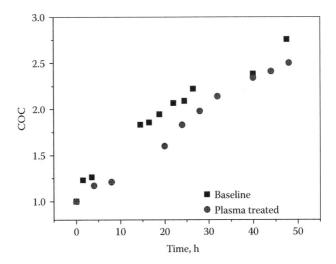

FIGURE 6.33
SEM photographs of the scales obtained for (a) no treatment and (b) plasma-treated cases ($CaCO_3$ hardness of 500 ppm and a flow velocity of 0.1 m/s). (From Yang, Y., Kim, H., Fridman, A., and Cho, Y.I. (2010) Effect of a plasma-assisted self-cleaning filter on the performance of PWT coil for the mitigation of mineral fouling in a heat exchanger. *Int. J. Heat Mass Transfer* 53, 412–422.)

size and above due to the precipitation effect of spark discharges, with blunt-edged crystals. The sharp and pointed crystal tips observed in the no treatment case are believed to be produced through precipitation reactions of mineral ions on the heat transfer surface, thus adhering to the heat transfer surface more strongly than blunt crystals observed in the plasma case.

Figure 6.33a and 6.33b show similar SEM images ($\times 200$, $\times 500$, and $\times 1,000$) of $CaCO_3$ scales for the no treatment and plasma-treated cases for 500-ppm hard water at a flow velocity of 0.1 m/s. The SEM images obtained from the no treatment case showed a more organized regular structure, which can be attributed to the aforementioned particulate fouling on the heat exchanger surface. For plasma treatment, the images showed random rounded particles with a less-organized structure. The size of the particles obtained from the plasma treatment case was slightly larger than those retrieved from the no treatment case.

Calcium carbonate is a crystalline substance that exists in three polymorphs: calcite, aragonite, and vaterite (Smith, 1986). Each polymorph has a unique crystallographic structure with a unique XRD spectrum that serves as its fingerprint. The present XRD analyses were conducted to determine the crystallographic phase of scale deposits so that the focus was on spectrum peaks and not on the intensity. Figure 34a shows the standard XRD spectra of the calcite phases of calcium carbonate as a reference, which has a prominent peak of intensity at $2q = 29.5°$.

Figures 6.34b and 6.34c present the results of the XRD analyses for the no treatment and plasma-treated cases at a water hardness of 500 ppm and

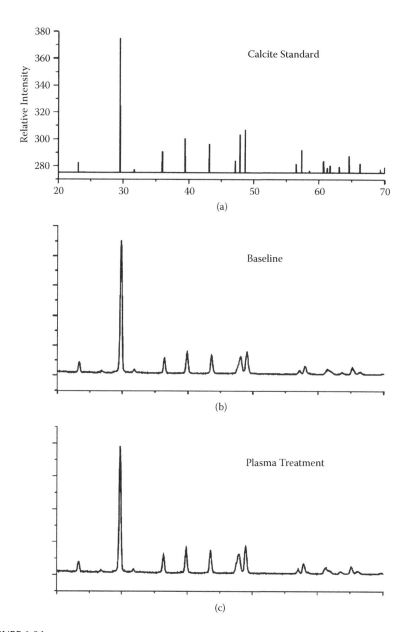

FIGURE 6.34
XRD analyses of the scales for (a) standard calcite; (b) no treatment; and (c) plasma-treated cases (500 ppm $CaCO_3$ hardness and a flow velocity of 0.1 m/s).

a flow velocity of 0.1 m/s. The results were compared to the standard XRD spectra of $CaCO_3$ given in Figure 6.34a. For both cases, the peaks depicted those of a calcite crystal. Although most previous studies reported aragonite crystals for the no treatment cases (Cho et al., 2004; Kontoyannis and Vagena, 1999), the present XRD result did not show aragonite for the no treatment case. This can be attributed to the fact that even for the no treatment cases there must have been a large number of suspended calcium particles in water in the present study due to extremely supersaturated states of cooling water and no blowdown in the study. Hence, even for the no treatment case, one might expect to have a large number of suspended particles in water, thus leading to a calcite form of $CaCO_3$ scales on the heat transfer surface.

For the cases of plasma treatment, spark discharge would produce more suspended calcium particles in hard water than the no treatment case (Yang, Kim, Sarikovskiy, et al., 2010), producing particulate fouling or a calcite form of calcium crystal at the heat transfer surface. The XRD results for the no treatment and plasma-treated cases at a water hardness of 250 and a flow velocity of 0.5 m/s were similar to the results given in Figure 6.34.

6.4.3 Cycle of Concentration

The COC is defined as the ratio of the dissolved solids in cooling tower water to those in makeup supply water. Figure 6.35 shows variations in COC over time for the case of 250-ppm hard water with a flow velocity of 0.5 m/s. The value of COC increased almost linearly with time because of zero blowdown. The COC reached approximately 2.8 at the end of the fouling test for

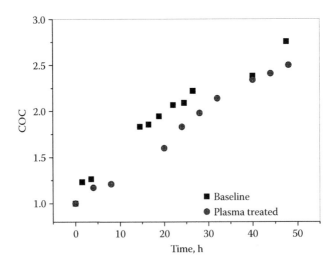

FIGURE 6.35
Variations in cycle of concentration (COC) versus time for 250 ppm hard water under no treatment and plasma-treated cases with a flow velocity of 0.5 m/s.

the no treatment case, whereas it arrived at 2.4 for the plasma-treated case. Since the hardness of the makeup water was 250 ppm, the hardness of circulating water became approximately 600–700 ppm at the end of the fouling tests for both cases. In addition, the COC value for the plasma-treated case was consistently smaller by about 0.3–0.5 than those for the no treatment cases during the entire fouling test, reflecting the fact that the pulsed spark discharge was continuously precipitating calcium ions from water. Note that the COC for other cases studied (i.e., at different flow velocities and different hardnesses) also reached approximately 3 at the end of fouling tests with zero blowdown, indicating that the water hardness was about 750 and 1,500 ppm for the 250- and 500-ppm cases, respectively, near the end of the test. Such extremely harsh fouling conditions were utilized in the study to expedite the fouling process and examine the performance and limitation of the pulsed spark discharge system.

References

Abou-Ghazala, A., Katsuki, S., Schoenbach, K.H., and Moreira, K.R. (2002) Bacterial decontamination of water by means of pulsed-corona discharges. *IEEE Trans. Plasma Sci.* 30, 1449–1453.

Adams, C., Timmons, T., Seitz, T., Lane, J., and Levotch, S. (2005) Trihalomethane and haloacetic acid disinfection by-products in full-scale drinking water systems. *J. Environ. Eng.* 131, 526–534.

Akishev, Y., Grushin, M., Karal'nik, V., Monich, A., Pan'kin, M., Trushkin, N., Kholodenko, V., Chugunov, V., Zhirkova, N., Irkhina, I., and Kobzev, E. (2006) Generation of a nonequlibrium plasma in heterophase atmospheric-pressure gas-liquid media and demonstration of its sterilization ability. *Plasma Phys. Rep.* 32(12), 1052–1061.

Akiyama, H. (2000) Streamer discharges in liquids and their applications. *Plasma Phys. Rep.* 7, 646–653.

Akiyama, H., Sakugawa, T., and Namihira, T. (2007) Industrial applications of pulsed power technology. *IEEE Trans. Dielectr. Electr. Insul.* 14, 1051–1064.

An, W., Baumung, K., and Bluhm, H. (2007) Underwater streamer propagation analyzed from detailed measurements of pressure release. *J. Appl. Phys.* 101, 053302.

Anpilov, A.M., Barkhudarov, E.M., Bark, Y.B., Zadiraka, Y.V., Christofi, N., Kozlov, Y.N., Kossyi, I.A., Silakov, V.P., Taktakishvili, M.I., and Temchin, S.M. (2001) Electric discharge in water as a source of UV radiation, ozone and hydrogen peroxide. *J. Phys. D: Appl. Phys.* 34, 993–999.

Anpilov, A.M., Barkhudarov, E.M., Christofi, N., Kopiev, V.A., Kossyi, I.A., Taktakishvili, M.I., and Zadiraka, Y. (2002) Pulsed high voltage electric discharge disinfection of microbially contaminated liquids. *Lett. Appl. Microbiol.* 35, 90–94.

Anpilov, A.M., Barkhudarov, E.M., Christofi, N., Kop'ev, V.A., Kossyi, I.A., Taktakishvili, M.I., and Zadiraka, Y.V. (2004) The effectiveness of a multi-spark electric discharge system in the destruction of microorganisms in domestic and industrial wastewaters. *J. Water Health* 2(4), 267–277.

Assael, M.J., Karagiannidis, L., Malamataris, N., and Wakeham, W.A. (1998) The transient hot-wire technique: a numerical approach. *Int. J. Thermophys.* 19, 379–389.

Avrorin, A. (1980) Shock compressibility at ~ 100 Mbar. *JETF Lett.*

Avrorin, A. (2006) Physical investigations under nuclear tests. UFN.

Baker, J.S., and Judd, S.J. (1996) Magnetic amelioration of scale formation. *Water Res.* 30, 247–260.

Baroch, R., Anita, V., Saito, N., and Takai, O. (2008) Bipolar pulsed electrical discharge for decomposition of organic compounds in water. *J. Electrostat.* 66, 294–299.

Beroual, A. (1993) Electronic and gaseous processes in the prebreakdown phenomena of dielectric liquids. *J. Appl. Phys.* 73, 4528–4533.

Beroual, A., Zahn, M., and Badent, A. (1998) Propagation and structure of streamers in liquid dielectrics. *IEEE Electr. Insul. Mag.* 14, 6–14.

Billamboz, N., Grivet, M., Foley, S., Baldacchino, G., and Hubinois, J.-C. (2010) Radiolysis of the polyethylene/water system: Studies on the role of hydroxyl radical. *Radiat. Phys. Chem.* 79, 36–40.

Bluhm, H., Frey, W., and Giese, H. (2000) Application of pulsed HV discharges to material fragmentation and recycling. *IEEE Trans. Dielectr. Electr. Insul.* 7, 625–636.

Bolorizadeh, M.A., and Rudd, M.E. (1986) Angular and energy dependence of cross sections for ejection of electrons from water vapor. I. 50–2000-eV electron impact. *Phys. Rev. A* 33, 880–887.

Bott, T.R. (1995) *The fouling of heat exchangers*, Elsevier Science, New York.

Bruggeman, P., and Leys, C. (2009) Non-thermal plasmas in and in contact with liquids. *J. Phys. D: Appl. Phys.* 42, 053001.

Bruggeman, P., Verreycken, T., González, M.Á., Walsh, J.L., Kong, M.G., Leys, C., and Schram, D.C. (2010) Optical emission spectroscopy as a diagnostic for plasmas in liquids: opportunities and pitfalls. *J. Phys. D: Appl. Phys.* 43(12), 124005.

Bruggeman, P., Walsh, J., Schram, D., Leys, C., and Kong, M.G. (2009) Time dependent optical emission spectroscopy of sub-microsecond pulsed plasmas in air with water cathode. *Plasma Sources Sci. Technol.* 18(4), 045023.

Buxton, G.V., and Elliot, A.J. (1986) Rate constant for reaction of hydroxyl radicals with bicarbonate ions. *Radiat. Phys. Chem.* 27, 241–243.

Carman, R.J., Mildren, R.P., Ward, B.K., and Kane, D.M. (2003) High-pressure (>1 bar) dielectric barrier discharge lamps generating short pulses of high peak power vaccum ultraviolet radiation. *J. Phys. D: Appl. Phys.* 37, 2399–2407.

Calvet, J. G. (1985) SO2, NO and NO2 oxidation mechanism: Atmospheric consideration. Butterworths-Heinemann, Oxford, UK.

Chalker, P.R., Bull, S.J., and Rickerby, D.S. (1991) A review of the methods for the evaluation of coating-substrate adhesion. *Mater. Sci. Eng. A* 140, 583–592.

Champion, C. (2003) Theoretical cross sections for electron collisions in water: structure of electron tracks. *Phys. Med. Biol.* 48, 2147–2168.

Chen, Y.-S., Zhang, X.-S., Dai, Y.-C., and Yuan, W.-K. (2004) Pulsed high-voltage discharge plasma for degradation of phenol in aqueous solution. *Sep. Purif. Technol.* 34, 5–12.

Chibowski, E., Holysz, L., and Wojcik, W. (1994) Changes in zeta potential and surface free energy of calcium carbonate due to exposure to radiofrequency electric field. *Colloids Surf., A* 92, 79–85.

Ching, W.K., Colussi, A.J., and Hoffmann, M.R. (2003) Soluble sunscreens fully protect *E. coli* from disinfection by electrohydraulic discharges. *Environ. Sci. Technol.* 37, 4901–4904.

Ching, W.K., Colussi, A.J., Sun, H.J., Nealson, K.H., and Hoffmann, M.R. (2001) *Escherichia coli* disinfection by electrohydraulic discharges. *Environ. Sci. Technol.* 35, 4139–4144.

Cho, Y.I., Fridman, A.F., Lee, S.H., and Kim, W.T. (2004) Physical water treatment for fouling prevention in heat exchangers. *Adv. Heat Transfer* 38, 1–72.

Cho, Y.I., Kim, W.T., and Cho, D.J. (2007) Electro-flocculation mechanism of physical water treatment for the mitigation of mineral fouling in heat exchangers. *Exp. Heat Transfer* 20, 323–335.

Cho, Y.I., Lane, J., and Kim, W. (2005) Pulsed-power treatment for physical water treatment. *Int. Commun. Heat Mass Transfer* 32, 861–871.

Cho, Y.I., Lee, S.H., and Kim, W.T. (2003) Physical water treatment for the mitigation of mineral fouling in cooling-water applications. *ASHRAE Trans.* 109, 346–357.

Choi, H.S., Shikovab, T.G., Titov, V.A., and Rybkin, V.V. (2006) Surface oxidation of polyethylene using an atmospheric pressure glow discharge with liquid electrolyte cathode. *J. Colloid Interface Sci.* 300, 640–647.

Chu, P.K., Tian, X.B., Wei, C.B., Yang, S.Q., and Fu, R.K.Y. (2006) Water plasma implantation/oxidation of magnesium alloys for corrosion resistance. *Nuclear Instrum. Methods Phys. Res., Sect. B* 242, 300–302.

Coetzee, P.P., Yacoby, M., Howell, S., and Mubenga, S. (1998) Scale reduction and scale modification effects induced by Zn and other metal species in physical water treatment. *Water SA* 24, 77–84.

Crittenden, J.C., Hu, S., Hand, D.W., and Green, S.A. (1999) A kinetic model for $H_2O_2/$ UV process in a completely mixed batch reactor. *Water Res.* 33, 2315–2328.

Czernichowski, A. (1994) Gliding arc: Applications to engineering and environment control. *Pure and Applied Chemistry* 66, 1301–1310.

Czernichowski, A. (1996) Gliding arc: Applications to engineering and environment control. *Pure Appl. Chem.* 66, 1301–1310.

Daniel, J., Jacob, J. (1986) Chemistry of OH in remote clouds and its role in the production of formic acid and peroxymonosulfate. J. Geophys. Res 91, 9807–9826.

De Baerdemaeker, F., Monte, M., and Leys, C. (2005) Capillary underwater discharges. *IEEE Trans. Plasma. Sci.* 33, 492–493.

De Baerdemaeker, F., Simek, M., and Leys, C. (2007) Efficiency of hydrogen peroxide production by ac capillary discharge in water solution. *J. Phys. D: Appl. Phys.* 40, 2801–2809.

De Baerdemaeker, F., Simek, M., Schmidt, J., and Leys, C. (2007) Characteristics of ac capillary discharge produced in electrically conductive water solution. *Plasma Sources Sci. Technol.* 16, 341–354.

Demadis, K.D., Mavredaki, E., Stathoulopoulou, A., Neofotistou, E., and Mantzaridis, C. (2007) Industrial water systems: problems, challenges and solutions for the process industries. *Desalination* 213, 38–46.

Deminsky, M. A., Potapkin, B.V., Rusanov, V. D., Fridman, A.A. (1990) Possibility of SO chain oxidation in heterogeneous air stream by relativistic electron beams. Kurchatov Institute of Atomic Energy, Moscow.

Denisov, E.T., Sarkisov, O.M., and Likhtenshtein, G.I. (2003) *Chemical kinetics fundamentals and new developments*, Elsevier, Amsterdam.

Derakhshesh, M., Abedi, J., and Hassanzadeh, H. (2010) Mechanism of methanol decomposition by non-thermal plasma. *J. Electrostat.* 68, 424–428.

Destaillats, H., Lesko, T.M., Knowlton, M., Wallac, H., and Hoffmann, M.R. (2001) Scale-up of sonochemical reactors for water treatment. *Ind. Eng. Chem. Res.* 40, 3855–3860.

Dhanasekaran, R., and Ramasamy, P. (1986) Two-dimensional nucleation in the presence of an electric field. *J. Crystal Growth* 79, 993–996.

Dingfelder, M., Hantke, D., Inokuti, M., and Paretzke, H.G. (1999) Electron inelastic-scattering cross sections in liquid water. *Radiat. Phys. Chem.* 53(1), 1–18.

Djuric, N.L., Cadez, I.M., and Kurepa, M.V. (1988) H2O and D2O total ionization cross-sections by electron impact. *Int. J. Mass Spectrom. Ion Processes* 83(3), R7–R10.

Donaldson, J., and Grimes, S. (1988) Lifting the scales from our pipes. *New Sci.* 18, 43–46.

Dors, M., Metel, E., and Mizeraczyk, J. (2007) Phenol degradation in water by pulsed streamer corona discharge and Fenton reaction. *Int. J. Plasma Environ. Sci. Technol.* 1, 76–81.

Dragsund, E., Andersen, A.B., and Johannessen, B.O. (2001) *Ballast water treatment by ozonation*, IMO, London.

Eggers, J. (1997) Nonlinear dynamics and breakup of free-surface flows. *Rev. Mod. Phys.* 69, 865–930.

Einaga, H., Ibusuki, T., and Futamura, S. (2001) Performance evaluation of a hybrid system comprising silent discharge plasma and manganese oxide catalysts for benzene decomposition. *IEEE Trans. Ind. Appl.* 37, 1476–1482.

El-Wakil, M.M. (1984) *Powerplant technology*, McGraw-Hill, New York.

Fathi, A., Mohamed, T., Gabrielli, C., Maurin, G., and Mohamed, B.A. (2006) Effect of a magnetic water treatment on homogeneous and heterogeneous precipitation of calcium carbonate. *Water Res.* 40, 1941–1950.

Feeley, T.J. (2008) DOE/Office of Fossil Energy's energy and water R&D program.

Feeley, T.J., Skone, T.J., Stiegel, G.J., Jr., McNemar, A., Nemeth, M., Schimmoller, B., Murphy, J.T., and Manfredo, L. (2008) Water: A critical resource in the thermoelectric power industry. *Energy* 33(1), 1–11.

Fridman, A. (2008) *Plasma chemistry*, Cambridge University Press, Cambridge.

Fridman, A., Gutsol, A., and Cho, Y.I. (2007) Non-thermal atmospheric pressure plasma. *Adv. Heat Transfer* 40, 1–134.

Fridman, A., and Kennedy, L. (2006) *Plasma physics and engineering*, Taylor & Francis, New York.

Fridman, G., Peddinghaus, M., Ayan, H., Fridman, A., Balasubramanian, M., Gutsol, A., Brooks, A., and Friedman, G. (2006) Blood coagulation and living tissue sterilization by floating electrode dielectric barrier discharge in air. *Plasma Chem. Plasma Proc.* 26, 425–442.

Gabrielli, C., Maurin, G., Francy-Chausson, H., Thery, P., Tran, T.T.M., and Tlili, M. (2006) Electrochemical water softening: principle and application. *Desalination* 206, 150–163.

Gao, J., Wang, X., Hu, Z., Deng, H., Hou, J., Lu, X., and Kang, J. (2003) Plasma degradation of dyes in water with contact glow discharge electrolysis. *Water Res.* 37(2), 267–272.

Gehr, R., Zhai, Z.A., Finch, J.A., and Rao, S.R. (1995) Reduction of soluble mineral concentration in $CaSO_4$ saturated water using a magnetic field. *Water Res.* 29, 933–940.

Ghezzar, M.R., Abdelmalek, F., Belhadj, M., Benderdouche, N., and Addou, A. (2007) Gliding arc plasma assisted photocatalytic degradation of anthraquinonic acid green 25 in solution with TiO_2. *Appl. Catal. B: Environ.* 72(3–4), 304–313.

Gidalevich, E., and Boxman, R. (2006) Sub- and supersonic expansion of an arc channel in liquid. *J. Phys. D: Appl. Phys.* 39, 652–659.

Gordillo-Vázquez, F.J. (2008) Air plasma kinetics under the influence of sprites. *J. Phys. D: Appl. Phys.* 41, 234016.

Grabowski, L., van Veldhuizen, E., Pemen, A., and Rutgers, W. (2006) Corona above water reactor for systematic study of aqueous phenol degradation. *Plasma Chem. Plasma Proc.* 26(1), 3–17.

Grahl, T., and Markl, H. (1996) Killing of microorganisms by pulsed electric fields. *Appl. Microbiol. Biotechnol.* 45, 148–157.

Grymonpre, D.R., Sharma, A.K., Finney, W.C., and Locke, B.R. (2001) The role of Fenton's reaction in aqueous phase pulsed streamer corona reactors. *Chem. Eng. J.* 82, 189–207.

Hagelaar, G. (2008) BOLSIG+ Solver. Laplace.

Handa, T., and Minamitani, Y. (2009) The effect of a water-droplet spray and gas discharge in water treatment by pulsed power. *IEEE Trans. Plasma Sci.* 37, 179–183.

Hayamizu, M., Tenma, T., and Mizuno, A. (1989) Destruction of yeast cells by pulsed high voltage application. *Proc. Inst. Electrostat. Jpn.* 13, 322.

He, F., and Hopke, P.K. (1993) Experimental study of ion-induced nucleation by radon decay. *J. Chem. Phys.* 99(12), 9972–9978.

Heesch, E.J.M., Pemen, A.J.M., Huijbrechts, A.H.J., van der Laan, P.C.T., Ptssinski, K.J., Zanstra, G.J., and de Jong, P. (2000) A fast pulsed power source applied to treatment of conducting liquids and air. *IEEE Trans. Plasma Sci.* 137, 137–140.

Herron, J.T., and Green, D.S. (2001) Chemical kinetics database and predictive schemes for nonthermal humid air plasma chemistry. Part II. Neutral species reactions. *Plasma Chem. Plasma Proc.* 21, 459–481.

Higashitani, K., and Oshitan, J. (1997) Measurements of magnetic effects on electrolyte by atomic force microscopy. *Trans. I ChemE B* 75, 115–119.

Hironori, A. (2008) Plasma generation inside externally supplied Ar bubbles in water. *Plasma Sources Sci. Technol.* 17(2), 025006.

Huang, J., Guo, W., and Xu, P. (2007) Comparative study of decomposition of CCl_4 in different atmosphere thermal plasmas. *Plasma Sci. Technol.* 9, 76–79.

Hurst, C.J. (2005) *Manual of environmental microbiology*, ASM Press, Washington, DC.

Incropera, F.P., Dewitt, D.P., Bergman, T.L., and Lavine, A.S. (2007) *Fundamentals of heat and mass transfer*, Wiley, New York.

Inoue, Y., and Kobayashi, T. (1993) Nonlinear oscillation of a gas-filled spherical cavity in an incompressible fluid. *Fluid Dyn. Res.* 11, 85–97.

Ishijima, T. (2010) Efficient production of microwave bubble plasma in water for plasma processing in liquid. *Plasma Sources Sci. Technol.* 19(1), 15010.

Ishijima, T., Hotta, H., Sugai, H., and Sato, M. (2007) Multibubble plasma production and solvent decomposition in water by slot-excited microwave discharge. *Appl. Phys. Lett.* 91(12), 121501–121503.

Itikawa, Y. (1974) Electron-impact vibrational excitation of H_2O. *J. Phys. Soc. Jpn.* 36, 1127–1132.

Itikawa, Y., and Mason, N. (2005) Cross Sections for electron collisions with water molecules. *J. Phys. Chem. Ref. Data* 34, 1–22.

Jasinski, M., Mieraczyk, J., Zakrzewski, Z., Ohkubo, T., Chang, J. S. (2002) CFC-11 destruction by microwave torch generated atmospheric-pressure nitrogen discharge. *J. Phys. D. Appl. Phys.* 35, 2274–2280.

Jensen, M.J., Bilodeau, R.C., Heber, O., Pedersen, H.B., Safvan, C.P., Urbain, X., Zajfman, D., and Andersen, L.H. (1999) Dissociative recombination and excitation of H_2O^+ and HDO^+ *Phys. Rev. A* 60, 2970–2976.

Joanna, P., and Satoshi, I. (2007) Removal of color caused by various chemical compounds using electrical discharges in a foaming column. *Plasma Processes Polym.* 4, 753–759.

Jones, H.M., and Kunhardt, E.E. (1994) Influence of pressure and conductivity on the pulsed breakdown of water. *Plasma Phys. Rep.*. 1, 1016–1025.

Jones, H.M., and Kunhardt, E.E. (1995) Development of pulsed dielectric breakdown in liquids. *J. Phys. D: Appl. Phys.* 28, 178–188.

Joseph, B.K., and Miksis, M. (1980) Bubble oscillations of large amplitude. *J. Acoust. Soc. Am.* 68, 628–633.

Joshi, A., Locke, B.R., Arce, P., and Finney, W.C. (1995) Formation of hydroxyl radicals, hydrogen peroxide and aqueous electrons by pulsed streamer corona discharge in aqueous solution. *J. Haz. Mater.* 41, 3–30.

Joshi, R., Qian, J., Zhao, G., Kolb, J.F., and Schoenbach, K.H. (2004) Are microbubbles necessary for the breakdown of liquid water subjected to a submicrosecond pulse? *J. Appl. Phys.* 96, 5129–5139.

Joshi, R.P., Hu, Q., Schoenbach, K.H., and Beebe, S.J. (2002) Simulations of electroporation dynamics and shape deformations in biological cells subjected to high voltage pulses. *IEEE Trans. Plasma Sci.* 30, 1536–1546.

Joshi, R.P., Kolb, J.F., Shu, X., and Schoenbach, K.H. (2009) Aspects of plasma in water: streamer physics and applications. *Plasma Processes Polym.* 6, 763–777.

Joshi, R.P., Qian, J., and Schoenbach, K.H. (2002) Electrical network-based time dependent model of electrical breakdown in water. *J. Appl. Phys.* 92, 6245–6251.

Jun, Q., Joshi, R.P., Schoenbach, K.H., Woodworth, J.R., and Sarkisov, G.S. (2006) Model analysis of self- and laser-triggered electrical breakdown of liquid water for pulsed-power applications. *IEEE Trans. Plasma Sci.* 34, 1680–1691.

Kalghatgi, S., Fridman, G., Cooper, M., Nagaraj, G., Peddinghaus, M., Balasubramanian, M., Vasilets, V., Gutsol, A., Fridman, A., and Friedman, G. (2007) Mechanism of blood coagulation by nonthermal atmospheric pressure dielectric barrier discharge plasma. *IEEE Trans. Plasma Sci.* 35, 1559–1566.

Kang, S. F., Liao, C. H., and Po, S. T. (2000) Decolorization of textile wastewater by photo-Fenton oxidation technology. *Chemosphere* 41, 1287–1297.

Katsuki, S., Akiyama, H., Abou-Ghazala, A., and Schoenbach, K.H. (2002) Parallel streamer discharge between wire and plane electrodes in water. *IEEE Trans. Dielect. Elect. Insul.* 9, 498–506.

Katz, J.L., Fisk, J.A., and Chakarov, V.M. (1994) Condensation of a supersaturated vapor IX. Nucleation on ions. *J. Chem. Phys.* 101(3), 2309–2318.

Kenyon, K.E. (1983) On the depth of wave influence. *J. Phys. Oceanogr.* 13, 1968–1970.

Kenyon, K.E. (1998) Capillary waves understood by an elementary method. *J. Oceanogr.* 54, 343–346.

Khare, S.P., and Meath, W.J. (1987) Cross sections for the direct and dissociative ionisation of NH_3, H_2O and H_2S by electron impact. *J. Phys. B: Atomic Mol. Phys.* 20(9), 2101.

Kim, W.T. (2001) *A study of physical water treatment methods for the mitigation of mineral fouling*, Drexel University, Philadelphia.

Kirkpatrick, M.J., Finney, W.C., and Locke, B.R. (2003) Chlorinated organic compound removal by gas phase pulsed streamer corona electrical discharge with reticulated vitreous carbon electrodes. *Plasma Polym.* 8, 165–177.

Kline, S.J., and McClintock, F.A. (1953) Describing uncertainties in single-sample experiments. *Mech. Eng.* 75, 3–8.

Kohchi, A., Adachi, S., and Nakagawa, Y. (1996) Decomposition of low-pressure pollutant by repeated pulse microwave discharge. *Jpn. J. Appl. Phys.* 35, 2326–2331.

Kolb, J., Joshi, R., Xiao, S., and Schoenbach, K. (2008) Streamers in water and other dielectric liquids. *J. Phys. D: Appl. Phys.* 41, 234007.

Kolikov, V.A., Kurochkin, V.E., Panina, L.K., and Rutberg, F.G. (2005) Pulsed electric discharges and prolonged microbial resistance of water. *Dokl. Biol. Sci.* 403, 279–281.

Kolikov, V.A., Kurochkin, V.E., Panina, L.K., Rutberg, A.F., Rutberg, F.G., Snetov, V.N., and Stogov, A.Y. (2007) Prolonged microbial resistance of water treated by a pulsed electrical discharge. *Tech. Phys.* 52(263–270).

Kong, M.G., Kroesen, G., Morfill, G., Nosenko, T., Shimizu, T., Dijk, J.v., and Zimmermann, J.L. (2009) Plasma medicine: an introductory review. *New J. Phys.* 11(11), 115012.

Kontoyannis, C.G., and Vagena, N.V. (1999) Calcium carbonate phase analysis using XRD and FT-Raman spectroscopy. *Analyst* 125, 251–255.

Korobeynikov, S.M., and Melekhov, A.V. (2002) *Microbubbles and breakdown initiation in water*, Graz, Austria.

Kostyuk, P.V. (2008) Effect of Ni and TiO 2 on hydrogen generation from aqueous solution with non-thermal plasma. *J. Phys. D: Appl. Phys.* 41(9), 095202.

Krcma, F., Z., S., and J., P. (2010) Diaphragm discharge in liquids: fundamentals and applications. *J. Phys.: Conf. Ser.* 207, 012010–012016.

Kupershtokh, A.L., and Medvedev, D.A. (2006) Anisotropic instability of dielectric liquids and decay to vapor-liquid system in strong electric fields. *Tech. Phys. Lett.* 32, 634–637.

Kushner, M.J. (1999) Strategies for rapidly developing plasma chemistry model. *Bull. Am. Phys. Soc.* 44, 63.

Kusic, H., Koprivanac, N., and Locke, B.R. (2005) Decomposition of phenol by hybrid gas/liquid electrical discharge reactors with zeolite catalysts. *J. Haz. Mater.* B125, 190–200.

Labas, M.D., Brandi, R.J., Martín, C.A., and Cassano, A.E. (2006) Kinetics of bacteria inactivation employing UV radiation under clear water conditions. *Chem. Eng. J.* 121, 135–145.

Lama, W., and Gallo, C. (1977) Systematic study of the electrical characteristics of the tricel current pulses from negative needle-to-plane coronas. *J. Appl. Phys.* 45, 103–113.

Lange, H., and Huczko, A. (2004) Carbon arc discharge: plasma emission spectroscopy and carbon nanostructure formation. *Trans. Mater. Res. Soc. Jpn.* 29, 3359–3364.

Langmuir, I. (1928) Oscillations in ionized gases. *Proc. Nat. Acad. Sci. U.S.* 14, 627–637.

Laroussi, M. (2005) Low temperature plasma-based sterilization: overview and state-of-the-art. *Plasma Process Polym.* 2, 391–400.

Laroussi, M. (2008) The biomedical applications of plasma: a brief history of the development of a new field of research. *IEEE Trans. Plasma Sci.* 36(4), 1612–1614.

Laroussi, M., Dobbs, F.C., Wei, Z., Doblin, M.A., Ball, L.G., Moreira, K.R., Dyer, F.F., and Richardson, J.P. (2002) Decontamination of water by excimer UV radiation. *IEEE Trans. Plasma Sci.* 30, 1501–1503.

Lee, G.J., Tijing, L.D., Pak, B.C., Baek, B.J., and Cho, Y.I. (2006) Use of catalytic materials for the mitigation of mineral fouling. *Int. Commun. Heat Mass Transfer* 33(1), 14–23.

Lee, H.Y., Uhm, H.S., Choi, H.N., Jung, Y.J., Kang, B.K., and Yoo, H.C. (2003) Underwater discharge and cell destruction by shockwaves. *J. Korean Phys. Soc.* 42, S880–S884.

Levine, I.N. (1978) *Physical chemistry*, McGraw-Hill, New York.

Lewis, T. (1994) Basic electrical processes in dielectric liquids. *IEEE Trans. Dielectr. Electr. Insul.* 1, 630–643.

Lewis, T. (1998) A new model for the primary process of electrical breakdown in liquids. *Plasma Phys. Rep..* 5, 306–315.

Lewis, T. (2003) Breakdown initiating mechanisms at electrode interfaces in liquids. *IEEE Trans. Dielect. Elect. Insul.* 10, 948–955.

Lezzi, A., and Prosperetti, A. (1987) Bubble dynamics in a compressible liquid. II. Second-order theory. *J. Fluid Mech.* 185, 289–321.

Li, J., Sato, M., and Ohshima, T. (2007) Degradation of phenol in water using a gas–liquid phase pulsed discharge plasma reactor. *Thin Solid Films* 515, 4283–4288.

Li, J., Zhou, Z., Wang, H., Li, G., and Wu, Y. (2007) Research on decoloration of dye wastewater by combination of pulsed discharge plasma and TiO_2 nanoparticles. *Desalination* 212, 123–128.

Lide, D.R. (ed.) (2005) *CRC handbook of chemistry and physics*, CRC Press, Boca Raton, FL.

Lin, L., Johnston, C.T., and Blatchley III, E.R. (1999) Inorganic fouling at quartz: water interfaces in ultraviolet photoreactors-I. chemical characterization. *Water Res.* 33, 3321–3329.

Lisitsyn, I.V., Nomiyama, H., Katsuki, S., and Akiyama, H. (1999a) Streamer discharge reactor for water treatment by pulsed power. *Rev. Sci. Instrum.* 70, 3457–3462.

Lisitsyn, I.V., Nomiyama, H., Katsuki, S., and Akiyama, H. (1999b) Thermal processes in a streamer discharge in water. *IEEE Trans. Dielectr. Electr. Insul.* 6, 351–356.

Liu, Y.J., and Jiang, X.Z. (2005) Phenol degradation by a nonpulsed diaphragm glow discharge in an aqueous solution. *Environ. Sci. Technol.* 39, 8512–8517.

Liu, Y.J., Jiang, X.Z., and Wang, L. (2007) One-step hydroxylation of benzene to phenol induced by glow discharge plasma in an aqueous solution. *Plasma Chem. Plasma Proc.* 27, 496–503.

Locke, B.R., Burlica, R., and Kirkpatrick, M.J. (2006) Formation of reactive species in gliding arc discharges with liquid water. *J. Electrostat.* 64, 35–43.

Locke, B.R., Sato, M., Sunka, P., Hoffmann, M.R., and Chang, J.S. (2006) Electrohydraulic discharge and nonthermal plasma for water treatment. *Ind. Eng. Chem. Res.* 45, 882–905.

Loeb, L.B. (1960) *Basic processes of gaseous electronics*, University of California Press, Berkeley.

Lu, X., Pan, Y., and Liu, K. (2002) Spark model of pulsed discharge in water. *J. Appl. Phys.* 91, 24–31.

Lukes, P., Clupek, M., Babicky, V., and Sunka, P. (2008) Pulsed electrical discharge in water generated using porous-ceramic-coated electrodes. *IEEE Trans. Plasma Sci.* 36, 1146–1147.

Lukes, P., Clupek, M., Babicky, V., and Sunka, P. (2009) The role of surface chemistry at ceramic/electrolyte interfaces in the generation of pulsed corona discharges in water using porous ceramic-coated rod electrodes. *Plasma Processes Polym.* 6, 719–728.

Lukes, P., and Locke, B.R. (2005) Degradation of substituted phenols in a hybrid gas-liquid electrical discharge reactor. *Ind. Eng. Chem. Res.* 44, 2921–2930.

Madigan, M.T., and Martinko, J.M. (2006) *Biology of microorganisms*, Prentice Hall, Upper Saddle River, NJ.

Magureanu, M., Piroi, D., Gherendi, F., Mandache, N., and Parvulescu, V. (2008) Decomposition of *Methylene Blue in Water by Corona Discharges. Plasma Chem. Plasma Proc.* 28(6), 677–688.

Magureanu, M., Piroi, D., Mandache, N.B., David, V., Medvedovici, A., and Parvulescu, V.I. (2010) Degradation of pharmaceutical compound pentoxifylline in water by non-thermal plasma treatment. *Water Res.* 44(11), 3445–3453.

Malik, M.A., Minamitani, Y., Xiao, S., Kolb, J.F., and Schoenbach, K.H. (2005) Streamers in water filled wire-cylinder and packed bed reactors. *IEEE Trans. Plasma Sci.* 33, 490–491.

Manolache, S., Somers, E.B., Wong, A.C.L., Shamamian, V., and Denes, F. (2001) Dense medium plasma environments: a new approach for the disinfection of water. *Environ. Sci. Technol.* 35, 3780–3785.

Manolache, S., Shamamian, V., Denes, F. (2004) DMP plasma-enhanced decontamination of water of aromatic compounds. *J. Environo. Eng.* 130, 17–25.

Marotta, E., Scorrano, G., and Paradisi, C. (2005) Ionic reactions of chlorinated volatile organic compounds in air plasma at atmospheric pressure. *Plasma Process Polym.* 2, 209–217.

Matsushima, Y., Yamazaki, T., Maeda, K., Noma, T., and Suzuki, T. (2006) Plasma oxidation of a titanium electrode in dc-plasma above the water surface. *J. Am. Ceram. Soc.* 89, 799–804.

Mededovic, S., Finney, W.C., and Locke, B.R. (2008) Electrical discharges in mixtures of light and heavy water. *Plasma Processes Polym.* 5, 76–83.

Meek, J.M., and Craggs, J.D. (1978) *Electrical breakdown of gases*, Wiley, New York.

Miichi, T., Ihara, S., Satoh, S., and Yamabe, C. (2000) Spectroscopic measurements of discharges inside bubbles in water. *Vacuum* 59(1), 236–243.

Millar, T.J., Farquhar, P.R.A., and Willacy, K. (1997) The UMIST Database for Astrochemistry 1995. *Astron. Astrophys. Suppl. Ser.* 121, 139–185.

Minamitani, Y., Shoji, S., Ohba, Y., and Higashiyama, Y. (2008) Decomposition of dye in water solution by pulsed power discharge in a water droplet spray. *IEEE Trans. Plasma Sci.* 36, 2586–2591.

Moisan, M., Barbeau, J., Crevier, M.-C., Pelletier, J., Philip, N., and Saoudi, B. (2002) Plasma sterilization. Methods and mechanisms. *Pure Appl. Chem.* 74(3), 349–358.

Moisan, M., Barbeau, J., Moreau, S., Pelletier, J., Tabrizian, M., and Yahia, L.H. (2001) Low temperature sterilization using gas plasmas: a review of the experiments and an analysis of the inactivation mechanisms. *Int. J. Pharm.* 226, 1–21.

Morch, K.A. (2007) Reflections on cavitation nuclei in water. *Phys. Fluids* 19, 072104–072110.

Muhammad Arif, M. (2009) Water purification by plasmas: which reactors are most energy efficient? *Plasma Chem. Plasma Proc.* 30(1), 21–31.

Muller-Steinhagen, H. (1999) Cooling water fouling in heat exchangers. *Adv. Heat Transfer* 33, 415–496.

Muller-Steinhagen, H. (2000) *Handbook of heat exchanger fouling—mitigation and cleaning technologies*, Publico, Germany.

Murphy, A.B. (1997) Destruction of ozone-depleting substances in a thermal plasma. *J. Proc. Roy. Soc. N. S. W.* 130, 93–108.

Murphy, A. B., Farmer, A. J. D., Horrigan, E. C., McAllister, T. (2002) Plasma destruction of ozone depleting substances. *Plasma Chem. Plasma Process* 22, 371–385.

Nagasaka, Y., and Nagashima, A. (1981) Simultaneous measurement of the thermal conductivity and the thermal diffusivity of liquids by the transient hot-wire method. *Rev. Sci. Instrum.* 52, 229–232.

Nakagawa, Y., Adachi, S., and Kohchi, A. (1996) Decomposition of chlorofluorocarbon by pulse high-current discharge and fast burning through spark discharge. *Jpn. J. Appl. Phys.* 35, 2808–2813.

Nantel-Valiquette, M., Kabouzi, Y., Castaños-Martinez, E., Makasheva, K., Moisan, M., and Rostaing, J.C. (2006) Reduction of perfluorinated compound emissions using atmospheric pressure microwave plasmas: Mechanisms and energy efficiency. *Pure Appl. Chem.* 78, 1173–1185.

Nicolae Bogdan, M. (2008) Decomposition of methylene blue in water using a dielectric barrier discharge: Optimization of the operating parameters. *J. Appl. Phys.* 104(10), 103306.

Nishioka, H., Saito, H., and Watanabe, T. (2009) Decomposition mechanism of organic compounds by DC water plasmas at atmospheric pressure. *Thin Solid Films* 518, 924–928.

Njatawidjaja, E., Sugiarto, A.T., Ohshima, T., and Sato, M. (2005) Decoloration of electrostatically atomized organic dye by the pulsed streamer corona discharge. *J. Electrostat.* 63, 353–359.

Nomura, S., and Toyota, H. (2003) Sonoplasma generated by a combination of ultrasonic waves and microwave irradiation. *Appl. Phys. Lett.* 83(22), 4503–4505.

Nomura, S., Toyota, H., Mukasa, S., Takahashi, Y., Maehara, T., Kawashima, A., and Yamashita, H. (2008) Discharge characteristics of microwave and high-frequency in-liquid plasma in water. *Appl. Phys. Express* 1, 046002.

Nomura, S., Toyota, H., Mukasa, S., Yamashita, H., Maehara, T., and Kuramoto, M. (2006) Microwave plasma in hydrocarbon liquids. *Appl. Phys. Lett.* 88(21), 211503–211503.

Ogata, A., Shintani, N., Yamanouchi, K., Mizuno, K., Kushiyama, S., and Yamamoto, T. (2000) Effect of water vapor on benzene decomposition using a nonthermal-discharge plasma reactor. *Plasma Chem. Plasma Proc.* 20, 453–467.

Olivero, J.J., Stagat, R.W., and Green, A. (1972) Electron deposition in water vapor, with atmospheric applications. *J. Geophys. Res.* 77, 4797–4811.

Opalska A., Opalinska T., and Ochman, P. (2002) Electroplasma-induced decomposition of chlorodifluoromethane under oxidizing conditions. *Acta Agrophys.* 80, 367–374.

Panchal, C.B., and Knudsen, J.G. (1998) Mitigation of water fouling: technology status and challenges. *Adv. Heat Transfer* 31, 431–474.

Parry, W., Bellows, J., Gallagher, J., and Harvey, A. (2008) *ASME international steam tables for industrial use*, American Society of Mechanical Engineers, New York.

Parsons, S.A., Judd, S.J., Stephenson, T., Udol, S., and Wang, B.L. (1997) Magnetically augmented water treatment. *Process Saf. Environ. Prot.* 75, 98–104.

Penetrante, B.M., Hsiao, M.C., Bardsley, J.N., Merritt, B.T., Vogtlin, G.E., and Wallman, P.H. (1996) Electron beam and pulsed corona processing of volatile organic compounds in gas streams. *Pure Appl. Chem.* 68, 1083–1087.

Penkett, S. A., Jones, B. M. R., Brice, K. A., Eggleton, A. E. J. (1979) The importance of atmospheric ozone and hydrogen peroxide in oxidizing sulphur dioxide in cloud and rain water. *Atmos. Environ.* 13, 123–137.

Perkins, W. S. (1999) Oxidative decolorization of dyes in aqueous medium. Text. Chem. Color 1, 33–37.

Probstein, R.F. (1989) *Physicochemical hydrodynamics*, Butterworths, Boston.

Prosperetti, A., and Lezzi, A. (1986) Bubble dynamics in a compressible liquid. I. First-order theory. *J. Fluid Mech.* 168, 457–478.

Qian, J., Joshi, R., Kolb, J.F., and Schoenbach, K.H. (2005) Microbubble-based model analysis of liquid breakdown initiation by a submicrosecond pulse. *J. Appl. Phys.* 97, 113304–113313.

Qian, J., Joshi, R., and Schoenbach, K.H. (2006) Model analysis of self- and laser-triggered electrical breakdown of liquid water for pulsed-power applications. *IEEE Trans. Plasma Sci.* 34, 1680–1691.

Quan, Z., Chen, Y., Ma, C., Wang, C., and Li, B. (2009) Experimental study on anti-fouling performance in a heat exchanger with low voltage electrolysis treatment. *Heat Transfer Eng.* 30, 181–188.

Radler, S., and Ousko-Oberhoffer, U. (2005) *Optimised heat exchanger management—achieving financial and environmental targets*, Klostor Irsee, Germany.

Raether, H. (1964) *Electron avalanches and breakdown in gases*, Butterworth, London.

Raizer, Y.P. (1997) *Gas discharge physics*, Springer, Berlin.

Ram, N.M., Christman, R.F., and Cantor, K.P. (1990) *Significance and treatment of volatile organic compounds in water supplies*, CRC Press, Chelsea, MI.

Rico, V., Hueso, J., Cotrino, J., and Gonzalez-Elipe, A. (2010) Evaluation of different dielectric barrier discharge plasma configurations as an alternative technology for green C1 chemistry in the carbon dioxide reforming of methane and the direct decomposition of methanol. *J. Phys. Chem. A* 114, 4009–4016.

Robinson, J.W., Ham, M., and Balaster, A.N. (1973) Ultraviolet radiation from electrical discharges in water. *J. Appl. Phys.* 44, 72–75.

Rosario-Ortiz, F.L., Mezyk, S.P., Doud, D., and Snyder, S.A. (2008) Quantitative correlation of absolute hydroxyl radical rate constants with non-isolated effluent organic matter bulk properties in water. *Environ. Sci. Technol.* 42, 5924–5930.

Sahni, M., and Locke, B.R. (2006) Quantification of reductive species produced by high voltage electrical discharges in water. *Plasma Processes Polym.* 3, 342–354.

Sakiyama, Y., Tomai, T., Miyano, M., and Graves, D.B. (2009) Disinfection of *E. coli* by nonthermal microplasma electrolysis in normal saline solution. *Appl. Phys. Lett.* 94(16).

Sato, K., and Yasuoka, K. (2008) Pulsed discharge development in oxygen, argon, and helium bubbles in water. *IEEE Trans. Plasma Sci.* 36, 1144–1145.

Sato, K., Yasuoka, K., and Ishii, S. (2008) Water treatment with pulsed plasmas generated inside bubbles. *Trans. Inst. Electr. Eng. Jpn., Part A* 128, 401–406.

Sato, K., Yasuoka, K., and Ishii, S. (2010) Water treatment with pulsed discharges generated inside bubbles. *Electr. Eng. Jpn.* (English translation of Denki Gakkai Ronbunshi) 170, 1–7.

Sato, M. (2008) Environmental and biotechnological applications of high-voltage pulsed discharges in water. *Plasma Sources Sci. Technol.* 17, 024021–024027.

Sato, M., Ohgiyama, T., and Clements, J.S. (1996) Formation of chemical species and their effects on microprganisms using a pulsed high-voltage discharge in water. *IEEE Trans. Ind. Appl.* 32, 106–112.

Sato, M., Tokutake, T., Ohshima, T., and Sugiarto, A.T. (2008) Aqueous phenol decomposition by pulsed discharges on the water surface. *IEEE Trans. Ind. Appl.* 44, 1397–1402.

Schmid, S., Jecklin, M., and Zenobi, R. (2010) Degradation of volatile organic compounds in a non-thermal plasma air purifier. *Chemosphere* 79, 124–130.

Schoenbach, K., Joshi, R., Kolb, J., Chen, N., Stacey, M., Blackmore, P.F., Buescher, E.S., and Beebe, S.J. (2004) Ultrashort electrical pulses open a new gateway into biological cells. *Proc. IEEE* 92(7), 1122–1137.

Schoenbach, K., Kolb, J., Shu, X., Katsuki, S., Minamitani, Y., and Joshi, R. (2008) Electrical breakdown of water in microgaps. *Plasma Sources Sci. Technol.* 17, 024010 (024010 pp.).

Schoenbach, K.H., Hargrave, B., Joshi, R.P., Kolb, J.F., Nuccitelli, R., Osgood, C., Pakhomov, A., Stacey, M., Swanson, R.J., White, J.A., Shu, X., Jue, Z., Beebe, S.J., Blackmore, P.F., and Buescher, E.S. (2007) Bioelectric effects of intense nanosecond pulses. *IEEE Trans. Dielectr. Electr. Insul.* 14, 1088–1109.

Schoenbach, K.H., Peterkin, F.E., Alden, R.W.J., and Beebe, S.J. (1997) The effect of pulsed electric fields on biological cells: experiments and applications. *IEEE Trans. Plasma Sci.* 25, 284–289.

Schutten, J., de Heer, F.J., Moustafa, H.R., Boerboom, A.J.H., and Kistemaker, J. (1966) Gross- and partial-ionization cross sections for electrons on water vapor in the energy range 0.1–20 keV. *J. Chem. Phys.* 44(10), 3924–3928.

Shen, Y., Lei, L., Zhang, X., Zhou, M., and Zhang, Y. (2008) Effect of various gases and chemical catalysts on phenol degradation pathways by pulsed electrical discharges. *J. Haz. Mater.* 150, 713–722.

Shin, W.-T., Yiacoumi, S., Tsouris, C., and Dai, S. (2000) A pulseless corona-discharge process for the oxidation of organic compounds in water. *Ind. Eng. Chem. Res.* 39(11), 4408–4414.

Siegal, F., and Reams, M.W. (1966) Temperature effect on precipitation of calcium carbonate from calcium bicarbonate solutions and its application to cavern environments. *Sedimentology* 7, 241–247.

Smith, C., Coetzee, P.P., and Meyer, J.P. (2003) The effectiveness of a magnetic physical water treatment device on scaling in domestic hot-water storage tanks. *Water SA* 29, 231–236.

Smith, W.F. (1986) *Principles of materials science and engineering*, McGraw-Hill, Singapore.

Snizhko, L.O., Yerokhin, A.L., Gurevina, N.L., Patalakha, V.A., and Matthews, A. (2007) Excessive oxygen evolution during plasma electrolytic oxidation of aluminium. *Thin Solid Films* 516, 460–464.

Snoeyink, V.L., and Jenkins, D. (1982) *Water chemistry*, Wiley, New York.

Somerscales, E.F.C. (1990) Fouling of heat exchangers: an historical review. *Heat Transfer Eng.* 11(1), 19–36.

Sridharan, K., Harrington, S.P., Johnson, A.K., Licht, J.R., Anderson, M.H., and Allen, T.R. (2007) Oxidation of plasma surface modified zirconium alloy in pressurized high temperature water. *Mater. Design* 28, 1177–1185.

Staack, D., Fridman, A., Gutsol, A., Gogotsi, Y., and Friedman, G. (2008) Nanoscale corona discharge in liquids, enabling nanosecond optical emission spectroscopy. *Angew. Chem. Int. Ed.* 47, 8020–8024.

Stalder, K.R., Mcmillen, D.F., and Woloszko, J. (2005) Electrosurgical plasmas. *J. Phys. D: Appl. Phys.* 38, 1728–1738.

Starikovskiy, A., Yang, Y., Cho, Y.I., and Fridman, A. (2011) Non-equilibrium plasma in liquid water: Dynamics of generation and quenching. *Plasma Sources Sci. Technol.* 20, 024003.

Straub, H.C., Lindsay, B.G., Smith, K.A., and Stebbings, R.F. (1998) Absolute partial cross sections for electron-impact ionization of H_2O and D_2O from threshold to 1000 eV. *J. Chem. Phys.* 108(1), 109–116.

Sugiarto, A.T., Ito, S., Ohshima, T., Sato, M., and Skalny, J.D. (2003) Oxidative decoloration of dyes by pulsed discharge plasma in water. *J. Electrostat.* 2003 58, 135–145.

Sugiarto, A.T., Ohshima, T., and Sato, M. (2002) Advanced oxidation processes using pulsed streamer corona discharge in water. *Thin Solid Films* 407, 174–178.

Sugiarto, A.T., and Sato, M. (2001) Pulsed plasma processing of organic compounds in aqueous solution. *Thin Solid Films* 386, 295–299.

Sun, B., Kunitomo, S., and Igarashi, C. (2006) Characteristics of ultraviolet light and radicals formed by pulsed discharge in water. *J. Phys. D: Appl. Phys.* 39(17), 3814–3820.

Sun, B., Sato, M., and Clements, J.S. (1997) Optical study of active species produced by a pulsed streamer corona discharge in water. *J. Electrostat.* 39, 189–132.

Sun, B., Sato, M., and Clements, J.S. (1999) Use of a pulsed high-voltage discharge for removal of organic compounds in aqueous solution. *J. Phys. D: Appl. Phys.* 32, 1908–1915.

Sun, B., Sato, M., Harano, A., and Clements, J.S. (1998) Non-uniform pulse discharge-induced radical production in distilled water. *J. Electrostat.* 43, 115–126.

Sung, T., Weng, W., Gu, A., and Jhan, X. (2010) Energy consumption of atmosphere pen-like plasma for decoloration. *Surface Coatings Technol.* 205, 5459–5461.

Sunka, P. (2001) Pulse electrical discharges in water and their applications. *Phys. Plasmas* 8, 2587–2594.

Sunka, P., Babicky, M., Clupek, M., Fuciman, M., Lukes, P., Simek, M., Benes, J., Locke, B., and Majcherova, Z. (2004) Potential applications of pulse electrical discharges in water. *Acta Phys. Slov.* 54, 135–145.

Sunka, P., Babicky, V., Clupek, M., Lukes, P., and Balcarova, J. (2003) Modified pin-hole discharge for water treatment. *Pulsed Power Conf. Digest Tech. Papers* 1, 229–231.

Sunka, P., Babicky, V., Clupek, M., Lukes, P., Simek, M., Schmidt, J., and Cernak, M. (1999) Generation of chemically active species by electrical discharges in water. *Plasma Sources Sci. Technol.* 8, 258–265.

Suzuki, T., Saburi, T., Tokunami, R., Murata, H., and Fujii, Y. (2006) Dominant species for oxidation of stainless steel surface in water vapor plasma. *Thin Solid Films* 506–507, 342–345.

Takeda, T., Jen-Shih, C., Ishizaki, T., Saito, N., and Takai, O. (2008) Morphology of high-frequency electrohydraulic discharge for liquid-solution plasmas. *IEEE Trans. Plasma Sci.* 36(4), 1158–1159.

Test, T. (2000) Plasma in water. *Plasma Phys. Rep..* 40, 125–136.

Tezuka, M., and Iwasaki, M. (1998) Plasma induced degradation of chlorophenols in an aqueous solution. *Thin Solid Films* 316, 123–127.

Tijing, L.D., Kim, H., Kim, C., Lee, D., and Cho, Y.I. (2009) Use of an oscillating electric field to mitigate mineral fouling in a heat exchanger. *Exp. Heat Transfer* 22, 257–270.

Tomizawa, S., and Tezuka, M. (2006) Oxidative degradation of aqueous cresols induced by gaseous plasma with contact glow discharge electrolysis. *Plasma Chem. Plasma Proc.* 26, 43–52.

Trevino, K., and Fisher, E. (2009) Detection limits and decomposition mechanisms for organic contaminants in water using optical emission spectroscopy. *Plasma Process Polym.* 6, 180–189.

Trunin, R.F. (1994) Shock compressibility of condensed materials in strong shock waves generated by underground nuclear explosions. *Phys. Uspekhi* 37, 1123.

U.S. Department of Energy (1998) *Non-chemical technologies for scale and hardness control.* DOE/EE-0162.

U.S. Environmental Protection Agency Office of Water. (2009) *National water quality inventory: report to Congress.* 2004 Reporting Cycle.

Ushakov, G.V. (2008) Antiscaling treatment of water by an electric field in heat-supply networks. *Thermal Eng.* 55, 570–573.

Vankov, A., and Palanker, D. (2007) Nanosecond plasma-mediated electrosurgery with elongated electrodes. *J. Appl. Phys.* 101(12).

Volenik, K., Nop, P., Kopoeiva, P., Kolman, B., and Dubsky, J. (2006) Isothermal oxidation of metallic coatings deposited by a water-stabilized plasma gun. *Metal. Mater.* 44, 40–48.

Wait, I.W. (2005) *Fouling of quartz surfaces in potable water ultraviolet disinfection reactors*, Purdue University, West Lafayette, IN.

Wang, Y.-F., Lee, W.-J., Chen, C.-Y., and Hsieh, L.-T. (1999) Reaction mechanisms in both a $CHF_3/O_2/Ar$ and $CHF_3/H_2/Ar$ radio frequency plasma environment. *Indus. Eng. Chem. Res.* 38(9), 3199–3210.

Wang, Z., Xu, D., Chen, Y., Hao, C., and Zhang, X. (2008) Plasma decoloration of dye using dielectric barrier discharges with earthed spraying water electrodes. *J. Electrostat.* 66, 476–481.

Watanabe, T., Taira, T., Takeuchi, A. (2005). CFC Destruction by steam plasmas generated by atmospheric DC discharge. 17th Int. Symposium on Plasma Chemistry, Toronto, Canada 1153–1158.

Watson, P., and Chadbank, W. (1991) The role of electrostatic and hydrodynamic forces in the negative-point breakdown of liquid dielectrics. *IEEE Trans. Electr. Insul.* 26, 543–559.

Weast, R., and David, R.L. (1986) *Handbook of chemistry and physics,* CRC Press, Boca Raton, FL.

White, F.M. (2006) *Viscous fluid flow*, McGraw-Hill, Singapore.

Wilson, M.P., Balmer, L., and Given, M.J. (2006) Application of electric spark generated high power ultrasound to recover ferrous and non-ferrous metals from slag waste. *Minerals Eng.* 19, 491–499.

Wolfe, R.L. (1990) Ultraviolet disinfection of potable water. *Environ. Sci. Technol.* 24, 768–773.

Woodworth, J.R., Lehr, J.M., and Elizondo-Decanini, J. (2004) Optical and pressure diagnostics of 4-MV water switches in the Z-20 test facility. *IEEE Trans. Plasma Sci.* 32, 1778–1789.

World Health Organization. (2010) *Progress on sanitation and drinking-water*.

World Water Assessment Programme. (2009) *World Water Assessment Programme. The 3rd United Nations world water development report: water in a changing world*, UNESCO, Paris.

Xiaokai, X. (2008) Research on the electromagnetic anti-fouling technology for heat transfer enhancement. *Appl. Therm. Eng.* 28, 889–894.

Xiaokai, X., Chongfang, M., and Yongchang, C. (2005) Investigation on the electromagnetic anti-fouling technology for scale prevention. *Chem. Eng. Technol.* 28, 1540–1545.

Xie, H., Gu, H., Fujii, M., and Zhang, X. (2006) Short hot wire technique for measuring thermal conductivity and thermal diffusivity of various materials. *Meas. Sci. Technol.* 17, 208–214.

Yamada, Y., Sun, B., Ohshima, T., Sato, M. (1998) Pulsed discharge in water through pinhole. Asia-Pacific Workshop on Water and Air Treatment by Advanced Oxidation Technologies: Innovation and Commerical Applications. Tsukubu, Japan.

Yan, J.H., Du, C., Li, X., Sun, X., Ni, M., Cen, K., and Cheron, B. (2005) Plasma chemical degradation of phenol in solution by gas–liquid gliding arc discharge. *Plasma Sources Sci. Technol.* 14, 637–644.

Yan, Z. C., Li, C., and Lin, W. H. (2009) Hydrogen generation by glow discharge plasma electrolysis of methanol solutions. *Inter. J. Hydrogen Energy* 34, 48–55.

Yang, H., Zhang, X., Wen, S., and Yuan, W. (2009) Decomposition of organic compounds in water by direct high voltage discharge. *Chem. Eng. Technol.* 32, 887–890.

Yang, Y., Gutsol, A., Fridman, A., and Cho, Y.I. (2009) Removal of $CaCO_3$ scales on a filter membrane using plasma discharge in water. *Int. J. Heat Mass Transfer* 52, 4901–4906.

Yang, Y., Kim, H., Fridman, A., and Cho, Y.I. (2010) Effect of a plasma-assisted self-cleaning filter on the performance of PWT coil for the mitigation of mineral fouling in a heat exchanger. *Int. J. Heat Mass Transfer* 53, 412–422.

Yang, Y., Kim, H., Starikovskiy, A., Fridman, A., and Cho, Y.I. (2010) Application of pulsed spark discharge for calcium carbonate precipitation in hard water. *Water Res.* 44, 3659–3668.

Yang, Y., Kim, H., Starikovskiy, A., Fridman, A., and Cho, Y.I. (2011a) Application of pulsed spark discharge for mineral fouling mitigation in a heat exchanger. *J. Heat Transfer* 133, 054502.

Yang, Y., Kim, H., Starikovskiy, A., Fridman, A., and Cho, Y.I. (2011b) Precipitation of calcium ions from hard water using pulsed spark discharges and its mechanism. *Plasma Chem. Plasma Proc.* 31, 51–66.

Yang, Y., Kim, H., Starikovskiy, A., Fridman, A., and Cho, Y.I. (2011c) Pulsed multi-channel discharge array in water with stacked circular disk electrodes. *IEEE Trans. Plasma Sci.* in press, DOI:10.1109/TPS.2011.2129600.

Yang, Y., Starikovskiy, A., Fridman, A., and Cho, Y.I. (2011) Analysis of streamer propagation for electric breakdown in liquid/bioliquid. *Plasma Med.* 1, 65–83.

Yong Cheol, H., Hyun Jae, P., Bong Ju, L., Won-Seok, K., and Han Sup, U. (2010) Plasma formation using a capillary discharge in water and its application to the sterilization of *E. coli. Phys. Plasmas* 17, 053502–053506.

Yukhymenko, V.V., Chernyak, V.Y., and Olshevskii, S.V. (2008) Plasma conversion of ethanol-water mixture to synthesis gas. *Ukr. J. Phys.* 53, 409–413.

Zhang, R., Zhang, C., Cheng, X., Wang, L., Wu, Y., and Guan, Z. (2007) Kinetics of decolorization of azo dye by bipolar pulsed barrier discharge in a three-phase discharge plasma reactor. *J. Haz. Mater.* 142, 105–110.

Zhang, R.B., Wang, L., Wu, Y., Guan, Z., and Jia, Z. (2006) Bacterial decontamination of water by bipolar pulsed discharge in a gas-liquid-solid three-phase discharge reactor. *IEEE Trans. Plasma Sci.* 34, 1370–1374.

Index

For Product Safety Concerns and Information please contact our EU
representative GPSR@taylorandfrancis.com Taylor & Francis Verlag GmbH,
Kaufingerstraße 24, 80331 München, Germany

Printed and bound by CPI Group (UK) Ltd, Croydon, CR0 4YY
02/05/2025
01859321-0002